Think Like a Coder: Student Edition

Screen-Free Activities That Teach Real Coding Skills

by

Lisa Hargreaves

Logic Lab Books

THIS BOOK BELONGS TO:

(write your name)

Think Like a Coder

Published by Logic Lab Books

Printed in the United States of America

First Edition

ISBN: 979-8-9967470-0-9

For more information, visit LogicLabBooks.com

Table of Contents

How To Use This Book

This workbook is designed to help you build real problem-solving skills through fun, screen-free activities. Each page introduces a small idea, gives you a chance to practice it, and builds confidence step by step.

What You'll Learn

Throughout this book, you will explore the same thinking skills real programmers use every day, including:

- **Patterns** — noticing what repeats
- **Logic** — making decisions based on rules
- **Sequencing** — putting steps in the right order
- **Debugging** — finding and fixing mistakes
- **Algorithms** — creating clear, step-by-step instructions

These skills form the foundation of coding, even without a computer.

How the Activities Work

Each activity is designed to be:

- **Short** — most take 5–10 minutes
- **Independent** — you can work on your own
- **Hands-on** — puzzles, patterns, and visual logic
- **Confidence-building** — each page builds on the last

There are no trick questions and no timed challenges. Go at your own pace.

For grown-ups: use the Parent Helper Boxes to guide discussion and celebrate progress.

Activity Sets

You'll see small icons at the beginning of each activity set. These show which type of thinking the activity focuses on:

Activity Set A
Learn the Idea

Activity Set B
Practice the Skill

Activity Set C
Apply the Concept

Activity Set D
Debug the Mistakes

Activity Set E
Create Your Own

You don't need to memorize these — they're simply a helpful guide for helpers and teachers.

Tips for Helpers

- Encourage kids to talk through their thinking
- If they get stuck, ask: "*What do you notice?*" or "*What changed?*"
- Celebrate effort, not perfection

This book is meant to feel like a puzzle adventure, not a test.

You're Ready to Begin!

Start with Chapter 1 and work your way through, or jump to any activity that looks fun. There's no wrong way to explore.

Have fun thinking like a coder!

Chapter 1: Sequencing

Order Matters

Introduction: Sequencing

Have you ever tried to follow someone's directions and gotten totally confused? Maybe they said, "Go clean your room," and they didn't tell you *how*. Do you pick up toys first? Make the bed? Put clothes away? Where do you even start? That's what computers feel like **all the time**.

Computers are super fast and super smart — only at the things you tell them to do. They don't guess. They don't fill in missing steps. They don't say, "Oh, I know what you meant." They follow instructions **exactly**, one step at a time. That's why **sequencing** is so important.

Sequencing means putting steps in the right order so something actually works. If the steps are out of order, missing, or too big, the whole thing falls apart.

Think about this:

- If you put your shoes on before your socks... things get weird, fast.
- If you try to eat cereal before you pour it... you'll still be hungry.
- And if you tell a computer "Make a sandwich," the computer will just sit there forever.

Why? Because "make a sandwich" is actually **lots** of tiny steps:

- Get the bread
- Open the bag
- Take out two slices
- Get the peanut butter
- Open the jar
- Pick up a knife
- Scoop the peanut butter
- Spread it on the bread
- ...and so on!

Humans think in big steps. Computers need **small** steps.

Sequencing helps us break big ideas into little steps that make sense — to us *and* to a computer.

In this chapter, you'll practice:

- Putting steps in the right order
- Finding steps that are missing
- Fixing steps that don't make sense
- Breaking big steps into smaller ones
- Thinking like a robot (it's fun, I promise)

By the end, you'll be able to write instructions so clear that even a computer could follow them.

Ready to become a sequencing master?

Let's go!

Mini-Lesson: What Is Sequencing?

Sequencing means putting steps in the right order so something works. Every recipe, every set of directions, every game rule, and every computer program uses sequencing.

When the steps are in the right order, everything makes sense. When the steps are in the wrong order… things get silly fast.

Think about brushing your teeth:

- You can't brush before you put toothpaste on the brush.
- You can't put toothpaste on the brush before you get the toothpaste.
- You can't get the toothpaste before you find it.

Each step depends on the step before it.

That's sequencing.

Why Sequencing Matters to Computers

Humans are great at guessing what someone *meant*. Computers are not.

If you tell a human, "Make a sandwich," they'll figure it out. If you tell a computer, "Make a sandwich," it will do nothing — because you didn't give it any steps.

Computers need instructions that are:

- **Clear**
- **Complete**
- **In the right order**
- **Broken into tiny steps**

A human thinks:

"Get ready for school."

A computer needs:

1. Wake up
2. Sit up
3. Put feet on floor
4. Stand up
5. Walk to closet
6. Open closet door
7. Pick out clothes
8. ...and so on!

Big steps become **lots of little steps**.

That's sequencing.

How Sequencing Helps You Think Like a Coder

When you learn sequencing, you learn how to:

- Break big tasks into smaller steps
- Put steps in the right order
- Notice when something is missing
- Explain your thinking clearly
- Write instructions a computer could follow

These are the same skills real programmers use every day.

And you're about to practice them — one fun activity at a time.

Activity Set A: Fix the Scrambled Steps

Activity A1 — Morning Mix-Up

These steps for getting ready in the morning are out of order. Rewrite them in the correct sequence.

1. Put on your shoes
2. Brush your teeth
3. Eat breakfast
4. Wake up
5. Get dressed

1. ____________________________
2. ____________________________
3. ____________________________
4. ____________________________
5. ____________________________

Helper Box		
Hint	*Common Mistake*	*What to Look For*
Which steps must happen before others?	Kids often put "eat breakfast" too early.	A clear beginning (wake up) and a clear ending (shoes on).

Activity A2 — The Scrambled Sandwich

These steps for making a sandwich are mixed up. Rewrite them in the correct sequence.

1. Eat the sandwich
2. Spread peanut butter on the bread
3. Get two slices of bread
4. Open the peanut butter jar
5. Put the bread on a plate

1. ____________________

2. ____________________

3. ____________________

4. ____________________

5. ____________________

Helper Box		
Hint	*Common Mistake*	*What to Look For*
Look for steps that require something to already be done.	Kids often forget that the jar must be opened before spreading.	Materials first, actions second.

Activity A3 — The Robot's Broken Instructions

A robot tried to follow these steps to water a plant. The steps are in the wrong order.

Rewrite them in the correct sequence.

1. Pour water on the plant
2. Fill the watering can
3. Walk to the plant
4. Turn on the faucet
5. Pick up the watering can

1. ____________________

2. ____________________

3. ____________________

4. ____________________

5. ____________________

Helper Box		
Hint	*Common Mistake*	*What to Look For*
Imagine acting out the steps.	Kids sometimes put "pour water" before the can is filled.	A logical flow from getting materials to completing the task.

Activity A4 — Packing for Vacation

You're getting ready for a family trip. The steps to pack your suitcase are mixed up. Rewrite them in the correct sequence.

1. Put your clothes into the suitcase
2. Zip up the suitcase
3. Choose the outfits you want to bring
4. Add your toothbrush and other toiletries
5. Place your shoes at the bottom of the suitcase
6. Grab your favorite stuffed animal

1. ______________________________

2. ______________________________

3. ______________________________

4. ______________________________

5. ______________________________

6. ______________________________

Helper Box		
Hint	*Common Mistake*	*What to Look For*
Think about what must be gathered before anything can go into the suitcase.	Kids often want to zip the suitcase too early.	Choosing items → placing items → closing the suitcase.

Activity A5 — Make a Peanut Butter & Jelly Sandwich

You want to make a sandwich for lunch. The steps are out of order. Rewrite them in the correct sequence.

1. Get out two slices of bread
2. Spread peanut butter on one slice
3. Spread jelly on the other slice
4. Put the two slices together
5. Cut the sandwich in half
6. Put the sandwich on a plate

1. ______________________________

2. ______________________________

3. ______________________________

4. ______________________________

5. ______________________________

6. ______________________________

Helper Box		
Hint	*Common Mistake*	*What to Look For*
Look for steps that require something to already be done.	Kids sometimes put the slices together before both spreads are added.	Materials first, actions second, finishing touches last.

Activity A6 — Get Ready for a Nature Walk

You're getting ready for a nature walk, and the steps are mixed up. Rewrite them in the correct sequence.

1. Put on your walking shoes.
2. Grab your hat and sunscreen.
3. Fill your water bottle.
4. Pack a small bag for collecting leaves or rocks.
5. Put your backpack by the door.
6. Check the weather before heading outside.

1. ______________________________
2. ______________________________
3. ______________________________
4. ______________________________
5. ______________________________
6. ______________________________

Helper Box		
Hint	*Common Mistake*	*What to Look For*
Which steps must happen before leaving for the walk?	Kids often forget to check the weather or pack supplies before heading out.	A clear order: dressing → packing → getting ready to go.

Activity A7 — Plant a Seed in a Cup

You're planting a seed for a science project, and the steps are out of order. Rewrite them in the correct sequence.

1. Put soil into the cup
2. Make a small hole with your finger
3. Drop the seed into the hole
4. Cover the seed with soil
5. Water the soil gently
6. Place the cup in a sunny spot
7. Write your name on the cup

1. ______________________________

2. ______________________________

3. ______________________________

4. ______________________________

5. ______________________________

6. ______________________________

7. ______________________________

Helper Box		
Hint	*Common Mistake*	*What to Look For*
Imagine doing the steps with your hands.	Kids sometimes water the soil before the seed is inside.	A sequence that prepares the cup, plants the seed, and finishes with care steps.

Activity A8 — Feed the Dog

You're helping take care of the family dog, and the steps are mixed up. Rewrite them in the correct sequence.

1. Scoop the right amount of dog food
2. Pour the food into the bowl
3. Check that the water bowl is full
4. Call the dog over
5. Wash your hands
6. Put the food scoop away
7. Make sure the feeding area is clean

1. ____________________
2. ____________________
3. ____________________
4. ____________________
5. ____________________
6. ____________________
7. ____________________

Helper Box		
Hint	*Common Mistake*	*What to Look For*
What must happen before the dog can eat?	Kids often forget to check the water bowl.	A sequence that prepares the food, feeds the dog, and finishes with cleanup.

Activity Set B: Missing Steps

Activity B1 — Getting Ready for Bed

You're getting ready for bed, and some of the steps are missing. Fill in the missing steps so the sequence makes sense.

1. Put on your pajamas.
2.
3.
4. Get into bed.
5. Turn off the light.

Helper Box		
Hint	*Common Mistake*	*What to Look For*
What do you do between putting on pajamas and getting into bed?	Skipping tooth brushing or forgetting a calming step.	Two reasonable steps such as brushing teeth, reading a book, or saying goodnight.

Activity B2 — Sorting Recyclables

You want to help clean up after snack time, and some of the steps are missing. Fill in the missing steps so the sequence makes sense.

1. Collect the empty containers from the table.
2.
3. Put paper items in the recycling bin.
4.
5. Put food scraps in the compost bin.
6. Wash your hands

Helper Box		
Hint	*Common Mistake*	*What to Look For*
What needs to happen before and after sorting the recyclables?	Forgetting to rinse containers or mixing trash with recycling.	Steps like checking labels, rinsing bottles, or putting items in the correct bin.

Activity B3 — Washing Your Hands

You need to wash your hands, and some of the steps are missing. Fill in the missing steps so the sequence makes sense.

1. Turn on the water.
2.
3.
4. Rinse the soap off your hands.
5. Dry your hands with a towel.

Helper Box		
Hint	*Common Mistake*	*What to Look For*
What do you do right after turning on the water?	Forgetting to use soap or skipping the scrubbing step.	Steps like putting soap on hands and rubbing them together.

Activity B4 — Packing a Lunchbox

You're packing your lunch for school, and some steps are missing. Fill in the missing steps so the sequence makes sense.

1. Put your lunchbox on the counter.
2.
3. Add your sandwich to the lunchbox.
4.
5. Put in a napkin.
6. Close the lunchbox.

Helper Box		
Hint	*Common Mistake*	*What to Look For*
Think about what else goes into a lunchbox besides the main food.	Forgetting to add a drink or a snack.	Steps like opening the lunchbox or adding fruit before closing the lunchbox.

Activity B5 — Getting Ready to Paint

You want to start a painting project, and some steps are missing. Fill in the missing steps so the sequence makes sense.

1. Put paper on the table.
2.
3.
4. Dip your brush into the paint.
5. Start painting your picture.
6. Put the brush back in the water cup.

Helper Box		
Hint	*Common Mistake*	*What to Look For*
What do you need to set up before dipping your brush in paint?	Kids often forget to get water or choose their colors first.	Steps like getting the paintbrush or opening the paint.

Activity B6 — Making a Fruit Salad

You're helping make a fruit salad, and some steps are missing. Fill in the missing steps so the sequence makes sense.

1. Wash your hands.
2.
3. Cut the fruit into small pieces.
4.
5. Mix everything in a bowl.
6. Put the bowl in the fridge.

Helper Box		
Hint	*Common Mistake*	*What to Look For*
What needs to happen before cutting the fruit?	Forgetting to wash the fruit or get a cutting board.	A clear order: preparation → action → cleanup

Activity B7 — Getting Ready for a Bike Ride

You're getting ready to go on a bike ride, and some steps are missing. Fill in the missing steps so the sequence makes sense.

1. Put on your helmet.
2.
3.
4. Check that your tires have enough air.
5.
6. Start riding your bike.

Helper Box		
Hint	*Common Mistake*	*What to Look For*
Think about safety and preparation before riding.	Kids may forget to check brakes or bring water.	Steps like putting on shoes, checking brakes, or grabbing a water bottle.

Activity B8 — Preparing for a Park Cleanup

You're helping get ready for a park cleanup, and some steps are missing. Fill in the missing steps so the sequence makes sense.

1. Gather gloves and trash bags.
2.
3. Fill water bottles for everyone.
4.
5. Pack snacks for volunteers.
6.
7. Head to the park to start cleaning.

Helper Box		
Hint	*Common Mistake*	*What to Look For*
What needs to happen before and after packing supplies?	Forgetting to gather water bottles.	Steps like checking the weather or organizing supplies for the group.

Activity Set C: Break Up Large Instructions

Activity C1 — Brushing Your Teeth

You need to brush your teeth, and the instruction is too big. Break it into smaller steps so a computer could follow it exactly.

1. ________________________________
2. ________________________________
3. ________________________________
4. ________________________________
5. ________________________________

Helper Box		
Hint	*Common Mistake*	*What to Look For*
What do you do first before the toothbrush even touches your teeth?	Skipping steps like getting toothpaste or rinsing.	A clear sequence from preparing the toothbrush to finishing and cleaning up.

Activity C2 — Feeding a Pet Fish

You need to feed your pet fish, and the instruction is too big. Break it into smaller steps so a computer could follow it exactly.

1. ______________________________

2. ______________________________

3. ______________________________

4. ______________________________

5. ______________________________

Helper Box		
Hint	*Common Mistake*	*What to Look For*
Think about what you do before and after giving the fish food.	Forgetting to check how much food to use.	Steps that include preparation, feeding, and cleanup.

Activity C3 — Making Your Bed

You need to make your bed, and the instruction is too big. Break it into smaller steps so a computer could follow it exactly.

1.
2.
3.
4.
5.

Helper Box		
Hint	*Common Mistake*	*What to Look For*
What needs to be straightened or arranged first?	Skipping steps like pulling up the sheets before placing pillows.	A sequence that organizes the bed from bottom layers to top.

Activity C4 — Packing Your Backpack for School

You need to pack your backpack, and the instruction is too big. Break it into smaller steps so a computer could follow it exactly.

1. ____________________

2. ____________________

3. ____________________

4. ____________________

5. ____________________

Helper Box		
Hint	*Common Mistake*	*What to Look For*
Think about what you need to gather before putting anything into the backpack.	Forgetting important items like homework or a water bottle.	Steps that include gathering materials, checking supplies, and packing.

Activity C5 — Getting Ready to Bake Cookies

You want to bake cookies, and the instruction is too big. Break it into smaller steps so a computer could follow it exactly.

1. ______________________________

2. ______________________________

3. ______________________________

4. ______________________________

5. ______________________________

Helper Box		
Hint	*Common Mistake*	*What to Look For*
What do you need to do before mixing ingredients and before baking?	Forgetting to preheat the oven or measure ingredients.	Steps like gathering tools, measuring ingredients, mixing, and baking.

Activity C6 — Preparing for a Park Cleanup

You want to get ready for a park cleanup, and the instruction is too big. Break it into smaller steps so a computer could follow it exactly.

1.
2.
3.
4.
5.

Helper Box		
Hint	*Common Mistake*	*What to Look For*
Think about what needs to be packed before leaving the house.	Forgetting essentials like napkins, water, or sunscreen.	Steps that include gathering food, supplies, and outdoor items.

Activity C7 — Planting a Garden Row

You want to plant a row in your garden, and the instruction is too big. Break it into smaller steps so a computer could follow it exactly.

1. ____
2. ____
3. ____
4. ____
5. ____
6. ____
7. ____

Helper Box		
Hint	*Common Mistake*	*What to Look For*
What must happen before seeds go into the ground?	Skipping soil preparation or spacing steps.	A sequence that includes preparing the soil, planting, watering, and cleanup.

Activity C8 — Getting Ready for a Sleepover

You want to get ready for a sleepover, and the instruction is too big. Break it into smaller steps so a computer could follow it exactly.

1.
2.
3.
4.
5.
6.
7.

Helper Box		
Hint	*Common Mistake*	*What to Look For*
Think about everything you need to bring and what order you gather it in.	Forgetting important items like pajamas or a toothbrush.	Steps that include choosing items, packing them, and checking the list.

Activity Set D: Debug the Steps

Activity D1 — Getting Dressed for School

You're getting dressed for school. One of the steps is wrong. Circle the step that is wrong, then rewrite the steps correctly.

1. Put on your socks.
2. Put on your shoes.
3. Put on your jacket.
4. Put on your shirt.
5. Grab your backpack.

1. ______________________________
2. ______________________________
3. ______________________________
4. ______________________________
5. ______________________________

Helper Box		
Hint	*Common Mistake*	*What to Look For*
What needs to be on before shoes?	Kids often put on shoes too early.	Clothing layers → outerwear and supplies.

Activity D2 — Making a Sandwich

One of these steps is wrong. Circle the wrong step, then rewrite the steps correctly.

1. Get two slices of bread.
2. Spread peanut butter on one slice.
3. Put the sandwich in your backpack.
4. Spread jelly on the other slice.
5. Put the slices together.

1. ____________________
2. ____________________
3. ____________________
4. ____________________
5. ____________________

Helper Box		
Hint	*Common Mistake*	*What to Look For*
What should happen *after* the sandwich is finished?	Kids may include a step that belongs at the very end of lunch prep.	A sequence that completes the sandwich before any packing happens.

Activity D3 — Watering a Plant

One step doesn't belong. Circle the wrong step, then rewrite the steps correctly.

1. Fill the watering can.
2. Pour water into the soil.
3. Put the plant in the refrigerator.
4. Check that the soil is damp.
5. Empty extra water from the watering can.

1. ______________________________

2. ______________________________

3. ______________________________

4. ______________________________

5. ______________________________

Helper Box		
Hint	*Common Mistake*	*What to Look For*
Which step doesn't make sense for taking care of a plant?	Choosing a step that is unusual and not actually incorrect.	A sequence that keeps the plant in its normal environment.

Activity D4 — Getting Ready for Bed

Two steps are wrong. Circle both wrong steps, then rewrite the steps correctly.

1. Put on your pajamas.
2. Brush your teeth.
3. Turn off the light.
4. Read a book.
5. Put your toothbrush away.
6. Get into bed.

1. ______________________________
2. ______________________________
3. ______________________________
4. ______________________________
5. ______________________________
6. ______________________________

Helper Box		
Hint	*Common Mistake*	*What to Look For*
Which steps happen too early or too late?	Kids may think "turn off the light" can happen anytime.	A sequence that finishes all tasks before lights out.

Activity D5 — Cleaning Your Room

Two steps are wrong. Circle both wrong steps, then rewrite the steps correctly.

1. Pick up toys from the floor.
2. Put dirty clothes in the hamper.
3. Make your bed.
4. Throw clean clothes in the trash.
5. Put books on the shelf.
6. Dump your trash can onto the floor.

1. ______________________________
2. ______________________________
3. ______________________________
4. ______________________________
5. ______________________________
6. ______________________________

Helper Box		
Hint	*Common Mistake*	*What to Look For*
Which steps don't match the goal of cleaning?	Kids may overlook a step that is obviously the opposite of cleaning.	A sequence that organizes, tidies, and finishes with a clean space.

Activity D6 — Getting Ready for Soccer Practice

Two steps are wrong. Circle both wrong steps, then rewrite the steps correctly.

1. Put on your shin guards.
2. Tie your cleats.
3. Fill your water bottle.
4. Put your soccer ball in the freezer.
5. Put on your practice jersey.
6. Grab your gear bag.

1. ____________________
2. ____________________
3. ____________________
4. ____________________
5. ____________________
6. ____________________

Helper Box		
Hint	*Common Mistake*	*What to Look For*
Which steps don't fit the activity or are in the wrong phase?	Kids may miss that clothing steps must come before equipment steps.	A sequence that prepares clothing first, then equipment, then leaving.

Activity D7 — Baking Cookies

Two steps are reversed. Circle the wrong steps, then rewrite the steps correctly.

1. Wash your hands.
2. Get out the ingredients.
3. Put the cookies in the oven.
4. Preheat the oven.
5. Mix the ingredients.
6. Put the cookie dough on the baking sheet.

1. ______________________________

2. ______________________________

3. ______________________________

4. ______________________________

5. ______________________________

6. ______________________________

Helper Box		
Hint	*Common Mistake*	*What to Look For*
Which steps happen in the wrong order for baking?	Kids may think "put cookies in the oven" is fine before preheating.	A sequence that prepares, mixes, preheats, and bakes in the correct order.

Activity D8 — Getting Ready for a Picnic

One step is reversed and one step is incomplete. Circle both wrong steps, then rewrite the steps correctly.

1. Pack sandwiches into the cooler.
2. Put sunscreen on after you get home.
3. Get plates, napkins, and cups.
4. Put "food" in the cooler.
5. Fill the water bottles.
6. Put everything in the car.

1. ______________________________

2. ______________________________

3. ______________________________

4. ______________________________

5. ______________________________

6. ______________________________

Helper Box		
Hint	*Common Mistake*	*What to Look For*
Which step belongs earlier, and which step doesn't give enough information?	Kids may not notice that vague steps can cause problems.	A sequence that includes clear, complete actions in the right order.

Activity Set E: Write Your Own Steps

Activity E1 — Brushing Your Hair

You need to brush your hair, from beginning to end. Write all the steps you need to follow.

Helper Box		
Hint	*Common Mistake*	*What to Look For*
What do you do before the brush touches your hair?	Skipping steps like getting the brush or removing tangles gently.	A sequence that starts with preparation and ends with finishing touches.

Activity E2 — Feeding the Dog

You need to feed the dog, from beginning to end. Write all the steps you need to follow.

Helper Box		
Hint	*Common Mistake*	*What to Look For*
What must happen before the dog can start eating?	Forgetting to check the water bowl or measure the food.	Steps that include preparation, feeding, and cleanup.

Activity E3 — Getting Ready for School

You need to get ready for school, from beginning to end. Write all the steps you need to follow.

Helper Box		
Hint	*Common Mistake*	*What to Look For*
Think about what you do first when you wake up.	Jumping straight to leaving the house.	A sequence that includes dressing, eating, and gathering supplies.

Activity E4 — Making a Fruit Smoothie

You want to make a fruit smoothie, from beginning to end. Write all the steps you need to follow.

Helper Box		
Hint	*Common Mistake*	*What to Look For*
What needs to be prepared before blending?	Forgetting to wash fruit or put the lid on the blender.	Steps that include preparation, blending, and cleanup.

Activity E5 — Setting Up a Board Game

You want to set up a board game, from beginning to end. Write all the steps you need to follow.

Helper Box		
Hint	*Common Mistake*	*What to Look For*
What do you need to do before anyone can start playing?	Skipping steps like sorting pieces or reading the instructions.	A sequence that prepares the board, pieces, and players.

Activity E6 — Making a Salad

You want to make a salad, from beginning to end. Write all the steps you need to follow.

Helper Box		
Hint	*Common Mistake*	*What to Look For*
What must happen before the vegetables go into the bowl?	Forgetting to wash or chop ingredients.	Steps that include washing, preparing, mixing, and serving.

Activity E7 — Getting Ready for a Camping Trip

You want to get ready for a camping trip, from beginning to end. Write all the steps you need to follow.

Helper Box		
Hint	*Common Mistake*	*What to Look For*
Think about what you need to pack and prepare before leaving home.	Leaving out safety items or forgetting essential gear.	A sequence that includes gathering supplies, packing, and final checks.

Activity E8 — Baking a Cake

You want to bake a cake, from beginning to end. Write all the steps you need to follow.

Helper Box		
Hint	*Common Mistake*	*What to Look For*
What must happen before mixing and before baking?	Skipping steps like preheating the oven or measuring ingredients.	A sequence that includes preparation, mixing, baking, and cooling.

CERTIFICATE OF ACHIEVEMENT

Step-by-Step Superstar

This certificate is awarded to ______________________________ for completing all activities in **Chapter 1: Sequencing** of *Think Like a Coder*.

___________________________ Helper Signature

___________________________ Date

LOGIC LAB BOOKS

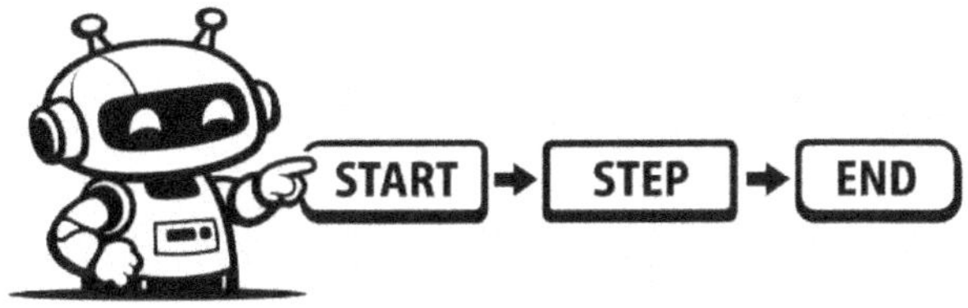

Chapter 2: Algorithms

Patterns with Purpose

Introduction: Algorithms

Have you ever tried to follow someone's directions and realized... they left out half the steps? Maybe they said, "Make your bed," and they didn't say *how*. Do you pull up the blanket first? Fix the pillows? Hunt for that one sheet that always disappears?

Humans can usually figure it out. Computers? Not a chance.

Computers are amazing — fast, powerful, and great at doing exactly what you tell them. And that's the catch: **exactly** what you tell them. Not what you *meant*. Not what you *probably wanted*. Only what you wrote down.

That's where **algorithms** come in.

An algorithm is a list of clear, precise steps that tells someone — or something — exactly how to do a task. You use algorithms all the time without even noticing:

- brushing your teeth
- tying your shoes
- making a snack
- getting ready for school

If you can explain how to do something step by step, congratulations — you already know how to write an algorithm.

Think about making a sandwich. If you tell a human, "Make a sandwich," they'll figure it out. If you tell a computer, "Make a sandwich," it will stare at you forever like, "Okay... *how*?"

A computer needs tiny steps:

1. Get two slices of bread
2. Open the peanut butter
3. Scoop some peanut butter
4. Spread it on the bread
5. Repeat for jelly
6. Put the slices together

Big ideas become small steps. That's the magic of algorithms.

Why Algorithms Matter Algorithms help computers understand exactly what to do — no guessing, no assumptions, no "I'll just wing it." When you learn to write algorithms, you learn to:

- think clearly
- explain your ideas
- break big tasks into smaller ones
- give instructions that *actually* work

These are the same skills real programmers use every day.

In this chapter, you'll practice turning big ideas into tiny steps, fixing instructions that don't work, comparing different algorithms, and writing your own.

Ready to think like a programmer? Let's dive in.

Mini-Lesson: What Is an Algorithm?

An algorithm is a step-by-step set of instructions that tells someone exactly how to do something. That's it. Simple, right?

Here's the important part: **Computers need algorithms because they can't guess.**

If you say, "Make a sandwich," a human fills in the missing pieces. A computer needs every step spelled out.

Good algorithms are:

- **clear** — easy to understand
- **complete** — no missing steps
- **in order** — steps happen at the right time
- **precise** — no guessing allowed

Let's look at another example: tying your shoes.

A human might say, "Tie your shoes." A computer needs:

1. Pick up the left lace
2. Pick up the right lace
3. Cross the laces
4. Tuck one end under the other
5. Pull tight
6. Make a loop
7. Wrap the other lace around the loop
8. Pull through
9. Tighten the bow

Suddenly, something simple becomes a whole list of steps.

That's what algorithms do — they turn everyday actions into clear instructions.

Why Algorithms Make You a Better Thinker

When you write algorithms, you learn to:

- slow down
- notice details
- explain your thinking
- break big ideas into smaller pieces
- communicate clearly

These are superpowers in programming — and in life.

In this chapter, you'll learn how to write, test, compare, and improve algorithms until they're so clear that even a computer could follow them perfectly.

Let's get started.

Activity Set A: Turn Big Ideas Into Algorithms

Activity A1 — Building a Paper Airplane

Turn this big idea into a clear algorithm a computer could follow.

1.
2.
3.
4.
5.

Helper Box		
Hint	*Common Mistake*	*What to Look For*
Think about what you need to start and finish the task.	Skipping steps like getting the paper or unfolding the wings.	A complete sequence from gathering supplies to testing.

Activity A2 — Washing Your Hands

Turn this big idea into a clear algorithm a computer could follow.

1.
2.
3.
4.
5.

Helper Box		
Hint	*Common Mistake*	*What to Look For*
What happens before the water touches your hands?	Forgetting to use soap or skipping the scrubbing step	A sequence that includes wetting, soaping, scrubbing, rinsing, and drying.

Activity A3 — Feeding a Pet

Turn this big idea into a clear algorithm a computer could follow.

1.

2.

3.

4.

5.

Helper Box		
Hint	*Common Mistake*	*What to Look For*
What must you prepare before the pet can eat?	Leaving out steps like measuring food or checking water.	A sequence that includes preparation, feeding, and cleanup.

Activity A4 — Packing a Backpack for School

Turn this big idea into a clear algorithm a computer could follow.

1.

2.

3.

4.

5.

6.

Helper Box		
Hint	*Common Mistake*	*What to Look For*
What do you need to gather before putting anything in the backpack?	Forgetting important items like homework or a water bottle.	A sequence that includes gathering, checking, and packing.

Activity A5 — Setting up a Bird Feeder

Turn this big idea into a clear algorithm a computer could follow.

1.
2.
3.
4.
5.
6.

Helper Box		
Hint	*Common Mistake*	*What to Look For*
What supplies do you need to collect?	Not finding a place to hang the bird feeder.	A sequence that includes gathering and preparing birdseed.

Activity A6 — Setting Up a Board Game

Turn this big idea into a clear algorithm a computer could follow.

1.

2.

3.

4.

5.

6.

Helper Box		
Hint	*Common Mistake*	*What to Look For*
What needs to be ready before anyone can start playing?	Skipping steps like sorting pieces or reading the instructions.	A sequence that prepares the board, pieces, and players.

Activity A7 — Organizing a Bookshelf

Turn this big idea into a clear algorithm a computer could follow.

1.
2.
3.
4.
5.
6.
7.

Helper Box		
Hint	*Common Mistake*	*What to Look For*
Think about what you need to do before you can start putting books in order.	Forgetting to remove the books before rearranging the shelves.	A sequence that includes removing, alphabetizing, and replacing books.

Activity A8 — Getting Ready for a Camping Trip

Turn this big idea into a clear algorithm a computer could follow.

1.

2.

3.

4.

5.

6.

7.

Helper Box		
Hint	*Common Mistake*	*What to Look For*
Think about what you need to pack and prepare before leaving home.	Leaving out safety items or forgetting essential gear.	A sequence that includes gathering supplies, packing, and final checks.

Activity Set B: Fix Vague Algorithms

Activity B1 — Folding a Paper Airplane

Turn this vague algorithm into a clearer one. Rewrite the steps so a computer could follow them exactly.

1. Fold the paper a few times.
2. Make the airplane wings.
3. Add the details.
4. Make it fly.

1. ____________________
2. ____________________
3. ____________________
4. ____________________
5. ____________________

Helper Box		
Hint	*Common Mistake*	*What to Look For*
Think about what folds must happen before the airplane shape appears.	Jumping straight to "make the airplane wings" without folding the body correctly.	Starting with basic folds, then wings, then details, then testing the flight.

Activity B2 — Setting Up a Robot Challenge

Turn this vague algorithm into a clearer one. Rewrite the steps so a computer could follow them exactly.

1. Test the robot's path.
2. Build the robot.
3. Program the robot's movement.
4. Test the robot's sensors.

1. ______________________________

2. ______________________________

3. ______________________________

4. ______________________________

5. ______________________________

Helper Box		
Hint	*Common Mistake*	*What to Look For*
What must happen before you can test the robot?	Starting the test before attaching sensors or finishing the program.	A sequence of building, connecting sensors, writing code, then testing.

Activity B3 — Cleaning Your Room

Turn this vague algorithm into a clearer one. Rewrite the steps so a computer could follow them exactly.

1. Clean up.
2. Put things away.
3. Be done.

1. ____________________
2. ____________________
3. ____________________
4. ____________________
5. ____________________

Helper Box		
Hint	*Common Mistake*	*What to Look For*
What kinds of things are on the floor, bed, or desk?	Writing "clean up" without saying what to clean or where it goes.	Steps that name specific items (toys, clothes, books) and where they should go.

Activity B4 — Sorting Recyclables

Turn this vague algorithm into a clearer one. Rewrite the steps so a computer could follow them exactly.

1. Gather stuff.
2. Sort it.
3. Toss it.

1. ________________________

2. ________________________

3. ________________________

4. ________________________

5. ________________________

Helper Box		
Hint	*Common Mistake*	*What to Look For*
What must happen before sorting items into bins?	Forgetting to check labels or mixing trash with recyclables.	A sequence gathering items, checking materials, then sorting correctly.

Activity B5 — Getting Ready for Soccer Practice

Turn this vague algorithm into a clearer one. Rewrite the steps so a computer could follow them exactly.

1. Get ready.
2. Go there.
3. Play.

1. ____________________

2. ____________________

3. ____________________

4. ____________________

5. ____________________

6. ____________________

Helper Box		
Hint	*Common Mistake*	*What to Look For*
What do you need to wear and bring before you can leave?	Leaving out equipment like cleats, shin guards, or a water bottle.	Steps that include dressing, packing gear, and traveling to practice.

Activity B6 — Preparing for a Nature Walk

Turn this vague algorithm into a clearer one. Rewrite the steps so a computer could follow them exactly.

1. Go outside.
2. Bring what you need.
3. Look for interesting things.
4. Get ready.

1. ____________________
2. ____________________
3. ____________________
4. ____________________
5. ____________________
6. ____________________
7. ____________________

Helper Box		
Hint	*Common Mistake*	*What to Look For*
What must happen before you go outside?	Skipping gathering the supplies before going outside.	A sequence including gathering, getting ready, and taking the walk.

Activity B7 — Planning a Community Garden Day

Turn this vague algorithm into a clearer one. Rewrite the steps so a computer could follow them exactly.

1. Get stuff.
2. Go there.
3. Have fun.

1.

2.

3.

4.

5.

6.

7.

Helper Box		
Hint	*Common Mistake*	*What to Look For*
What supplies and steps do you need before starting the garden work?	Forgetting to bring tools or skipping planting before watering.	Steps that include gathering supplies, traveling, planting, and caring for the garden.

Activity B8 — Hosting a Movie Night

Turn this vague algorithm into a clearer one. Rewrite the steps so a computer could follow them exactly.

1. Get ready.
2. Watch the movie.
3. Be done.

1. ____

2. ____

3. ____

4. ____

5. ____

6. ____

7. ____

Helper Box		
Hint	*Common Mistake*	*What to Look For*
What needs to be prepared before anyone presses play?	Forgetting steps like choosing the movie, or getting snacks.	A sequence including choosing a movie, and getting snacks ready.

Activity Set C: Compare Two Algorithms

Activity C1 — Making a Sandwich

Two algorithms are shown below. Choose the one that works better and explain why.

Algorithm A

1. Get two slices of bread.
2. Spread peanut butter on one slice.
3. Spread jelly on the other slice.
4. Put the slices together.

Algorithm B

1. Get two slices of bread.
2. Put the slices together.
3. Spread peanut butter and jelly on the outside.

Helper Box		
Hint	*Common Mistake*	*What to Look For*
Which algorithm puts the ingredients in the right place?	Choosing the algorithm with fewer steps just because it looks simpler.	An explanation that focuses on correct order and correct placement of ingredients.

Activity C2 — Brushing Your Teeth

Two algorithms are shown below. Choose the one that works better and explain why.

Algorithm A

1. Brush your teeth.
2. Put toothpaste on the toothbrush.
3. Rinse your mouth.

Algorithm B

1. Put toothpaste on the toothbrush.
2. Brush your teeth.
3. Rinse your mouth.

Helper Box		
Hint	*Common Mistake*	*What to Look For*
Which algorithm has the steps in a logical order?	Not noticing when a step happens too late to make sense.	An explanation that shows awareness of preparation before action.

Activity C3 — Feeding a Pet Fish

Two algorithms are shown below. Choose the one that works better and explain why.

Algorithm A

1. Sprinkle food into the tank.
2. Open the container of fish food.
3. Close the container.

Algorithm B

1. Open the container of fish food.
2. Sprinkle food into the tank.
3. Close the container.

____

____

____

Helper Box		
Hint	*Common Mistake*	*What to Look For*
Which algorithm avoids doing things out of order?	Ignoring steps that must happen before others.	An explanation that highlights the importance of preparation and cleanup.

Activity C4 — Packing a Backpack

Two algorithms are shown below. Choose the one that works better and explain why.

Algorithm A

1. Put your homework in the backpack.
2. Add your water bottle.
3. Zip the backpack.

Algorithm B

1. Put your homework in the backpack.
2. Zip the backpack.
3. Add your water bottle.

Helper Box		
Hint	*Common Mistake*	*What to Look For*
Which algorithm finishes with the backpack fully closed?	Choosing the algorithm that "looks" neat instead of the one that works.	An explanation that recognizes the importance of zipping last.

Activity C5 — Cleaning a Fish Tank

Two algorithms are shown below. Choose the one that works better and explain why.

Algorithm A

1. Gather the cleaning tools.
2. Move the fish to a temporary bowl.
3. Empty part of the tank water.
4. Clean the tank walls.
5. Add fresh water and return the fish.

Algorithm B

1. Clean the tank walls.
2. Add fresh water.
3. Remove the fish.
4. Empty part of the tank water.
5. Gather the cleaning tools.

Helper Box		
Hint	*Common Mistake*	*What to Look For*
Which steps must happen before you start cleaning the tank?	Not noticing that the fish must be moved before cleaning begins.	A logical order: gather tools → move fish → clean → refill.

Activity C6 — Setting Up a Board Game

Two algorithms are shown below. Choose the one that works better and explain why.

Algorithm A

1. Read the instructions.
2. Set up the board the right way.
3. Give each player their pieces.

Algorithm B

1. Put all the pieces on the board randomly.
2. Read the instructions.
3. Set up the board the right way.

____

____

____

Helper Box		
Hint	*Common Mistake*	*What to Look For*
Which algorithm starts with understanding the rules?	Choosing the algorithm that starts "doing things" instead of learning first.	An explanation that values reading instructions before setting up.

Activity C7 — Setting Up a Science Notebook

Two algorithms are shown below. Choose the one that works better and explain why.

Algorithm A

1. Decorate the cover.
2. Write today's date.
3. Number all the pages.
4. Add section labels.
5. Create a table of contents.

Algorithm B

1. Number all the pages.
2. Add section labels.
3. Create a table of contents.
4. Write today's date.
5. Decorate the cover.

Helper Box		
Hint	*Common Mistake*	*What to Look For*
Which steps should happen before you start writing or decorating?	Choosing the algorithm that looks more fun.	Organizing the notebook first (numbering, sections) before adding details.

Activity C8 — Getting Ready for a Camping Trip

Two algorithms are shown below. Choose the one that works better and explain why.

Algorithm A

1. Drive to the campsite.
2. Pack your tent, food, and sleeping bag.
3. Check the weather.
4. Load everything into the car.

Algorithm B

1. Check the weather.
2. Pack your tent, food, and sleeping bag.
3. Load everything into the car.
4. Drive to the campsite.

Helper Box		
Hint	*Common Mistake*	*What to Look For*
Which algorithm starts with planning before packing?	Choosing the algorithm that looks shorter instead of the one that makes sense.	An explanation that shows awareness of planning → packing → traveling.

Activity Set D: Write Your Own Algorithm

Activity D1 — Building a Domino Chain Reaction

Write a complete algorithm someone else could follow to build a domino chain.

Helper Box		
Hint	*Common Mistake*	*What to Look For*
Think about what must happen before you can start placing dominos in a line.	Starting to build the chain before choosing a flat surface.	Preparing the surface, spacing dominos evenly, and testing the chain at the end.

Activity D2 — Planting a Garden Row

Write a complete algorithm someone else could follow to plant a row of seeds.

Helper Box		
Hint	*Common Mistake*	*What to Look For*
What needs to be ready before the seeds go into the soil?	Forgetting steps like spacing seeds or watering afterward.	A sequence that includes preparing soil, planting, covering, and watering.

Activity D3 — Hosting a Birthday Party

Write a complete algorithm someone else could follow to prepare for a birthday party.

Helper Box		
Hint	*Common Mistake*	*What to Look For*
Think about planning, preparing, and setting up before guests arrive.	Jumping straight to the party without preparation steps.	A sequence including planning, decorating, preparing, and welcoming.

Activity D4 — Making a Video for School

Write a complete algorithm someone else could follow to create a short school video.

Helper Box		
Hint	*Common Mistake*	*What to Look For*
What needs to happen before recording begins?	Skipping steps like planning the script or checking equipment.	A sequence that includes planning, recording, reviewing, and finalizing.

Activity D5 — Preparing a Science Experiment

Write a complete algorithm someone else could follow to set up a simple science experiment.

Helper Box		
Hint	*Common Mistake*	*What to Look For*
What materials do you need before starting?	Leaving out safety steps or cleanup.	Gathering materials, preparing, performing, and cleaning up.

Activity D6 — Planning a Sleepover

Write a complete algorithm someone else could follow to get ready for a sleepover.

Helper Box		
Hint	*Common Mistake*	*What to Look For*
Think about what needs to be packed and prepared before leaving home.	Forgetting essentials like pajamas, toothbrush, or a sleeping bag.	A sequence that includes planning, packing, and final checks.

Activity D7 — Building a LEGO Structure

Write a complete algorithm someone else could follow to build a small LEGO structure.

Helper Box		
Hint	*Common Mistake*	*What to Look For*
What needs to happen before you start connecting pieces?	Skipping steps like sorting pieces or planning the design.	A sequence that includes planning, sorting, building, and adjusting.

Activity D8 — Preparing for a Camping Trip

Write a complete algorithm someone else could follow to prepare for a camping trip.

Helper Box		
Hint	*Common Mistake*	*What to Look For*
Think about planning, packing, safety, and travel.	Leaving out essential gear or skipping safety checks.	A sequence including planning, gathering supplies, packing, and checking.

CERTIFICATE OF ACHIEVEMENT

Algorithm Architect

This certificate is awarded to ______________________________ for completing all activities in **Chapter 2: Algorithms** of *Think Like a Coder*.

___________________________ Helper Signature

___________________________ Date

LOGIC LAB BOOKS

Chapter 3: Debugging

Every Error is a Clue

Introduction: Debugging

Have you ever followed directions *exactly* and still ended up with something totally wrong? Maybe you baked cookies and forgot the sugar. Maybe you built a LEGO set and realized you skipped a step 20 pages ago. Maybe you tried to follow a friend's instructions and ended up with... chaos. Congratulations — you've already experienced **debugging**.

Debugging is the process of finding and fixing mistakes in an algorithm so it works the way you expect. It's like being a detective: you look for clues, test ideas, and figure out what went wrong.

Computers don't guess. If there's a mistake in the instructions, the computer will do the wrong thing *perfectly*. If a step is missing, out of order, or just plain silly, the computer won't fix it — **you** have to.

That's why debugging matters. Think about this: If you tell a computer:

1. Put bread on plate
2. Spread peanut butter
3. Put bread on top

...and you forget to say "Get the bread," the computer won't magically know to grab it. It will try to spread peanut butter on... nothing. Messy, messy, messy.

Debugging helps you:

- spot mistakes
- understand why they happened
- fix them
- prevent them next time

Real programmers debug all the time — sometimes more than they write new code!

And here's the best part: **debugging isn't random guessing — it's a repeatable process.** In this book, you'll learn the **I CAN DO IT** method, a debugging loop that helps you stay calm, curious, and in control whenever something goes wrong.

I CAN DO IT stands for:

- **I — Investigate:** Look closely at what happened.
- **C — Contain:** Find the exact spot where the problem begins.
- **A — Analyze:** Ask "Why did this happen?"
- **N — Nudge:** Try one small change.
- **D — Debug:** Run the algorithm again with your small change.
- **O — Observe:** See what changed.
- **I — Iterate:** Try again if needed.
- **T — Teach Future You:** Write down what you learned so you don't repeat the mistake.

This loop turns mistakes into clues — and you into a real Debugging Detective.

In this chapter, you'll learn how to find errors, fix broken instructions, compare expected vs. actual results, and use the **I CAN DO IT** method to think like a real problem solver.

Ready to grab your detective hat? Let's debug.

Mini-Lesson: What Is Debugging?

Debugging means finding mistakes in an algorithm and fixing them so everything works the way it should.

Mistakes can be sneaky. Sometimes the problem is:

- a missing step
- a step in the wrong place
- a step that doesn't make sense
- a step that causes something unexpected

Debugging is how you track those mistakes down.

Imagine you're brushing your teeth and your steps say:

1. Put toothpaste on brush
2. Brush teeth
3. Turn on water
4. Rinse brush

Something's off, right? You turned on the water *after* brushing. Debugging helps you spot that and fix it.

Good debuggers:

- read carefully
- compare what *should* happen to what *did* happen
- look for patterns
- test one idea at a time
- stay calm and curious

The way you stay calm and curious is by using your debugging loop: **I CAN DO IT**.

When something goes wrong, you'll:

- **Investigate** the mistake,
- **Contain** the problem to the exact step,
- **Analyze** what caused it,
- **Nudge** the algorithm with a tiny change,
- **Debug** by trying the algorithm again with that tiny change,
- **Observe** the new result,
- **Iterate** until it works, and
- **Teach Future You** what you learned.

Debugging isn't about being perfect — it's about being a problem solver.

In this chapter, you'll practice debugging all kinds of algorithms, from tiny mistakes to tricky ones. By the end, you'll be able to fix broken instructions like a pro — and you'll have a reliable method you can use for the rest of your life.

Let's get debugging.

Activity Set A: Find the Mistake

Activity A1 — Making Toast

Find the mistake in this algorithm.

1. Put bread in the toaster.
2. Spread butter on the bread.
3. Push the toaster lever down.
4. Take out the toast when it pops.

Rewrite the algorithm so it makes sense.

Helper Box		
Hint	*Common Mistake*	*What to Look For*
Which step happens too early?	Not noticing when an action happens before the bread is toasted.	An explanation that identifies the step that must happen after toasting.

Activity A2 — Watering a Plant

Find the mistake in this algorithm.

1. Fill the watering can.
2. Water the plant.
3. Put the watering can away.
4. Turn on the faucet.

Rewrite the algorithm so it makes sense.

Helper Box		
Hint	*Common Mistake*	*What to Look For*
Which step is out of order?	Missing that the faucet should be turned on before filling the can.	A clear identification of the misplaced step.

Activity A3 — Getting Dressed

Find the mistake in this algorithm.

1. Put on your socks.
2. Put on your shoes.
3. Put on your shirt.
4. Tie your shoes.

Rewrite the algorithm so it makes sense.

Helper Box		
Hint	*Common Mistake*	*What to Look For*
Which step doesn't belong in the middle of getting dressed?	Not noticing that tying shoes must happen after both shoes are on.	A step that is correct but in the wrong place.

Activity A4 — Setting Up a Birdwatching Station

Find the mistake in this algorithm.

1. Put birdseed in the feeder.
2. Hang the feeder on a tree branch.
3. Choose a good spot to watch birds.
4. Fill the binoculars with batteries.

Rewrite the algorithm so it makes sense.

Helper Box		
Hint	*Common Mistake*	*What to Look For*
Which step doesn't make sense for the tools being used?	Not noticing that binoculars don't use batteries.	An explanation that identifies the step that doesn't belong in the process at all.

Activity A5 — Packing a Lunch

Find the mistake in this algorithm.

1. Put a sandwich in the lunchbox.
2. Add a drink.
3. Close the lunchbox.
4. Add a snack.

Rewrite the algorithm so it makes sense.

Helper Box		
Hint	*Common Mistake*	*What to Look For*
Which step happens after the lunchbox is already closed?	Not noticing that closing the lunchbox should be last.	A step that must move to the end.

Activity A6 — Cleaning a Room

Find the mistake in this algorithm.

1. Pick up toys from the floor.
2. Put dirty clothes in the hamper.
3. Vacuum the floor.
4. Dump out the toy bin.

Rewrite the algorithm so it makes sense.

Helper Box		
Hint	*Common Mistake*	*What to Look For*
Which step undoes the cleaning?	Not noticing that dumping toys makes the room messy again.	A step that should not be in the algorithm at all.

Activity A7 — Setting Up a Homework Caddy

Find the mistake in this algorithm.

1. Put your pencils and markers into the caddy.
2. Throw all your homework papers into the trash.
3. Add a ruler, scissors, and glue stick.
4. Place the caddy on your desk.

Rewrite the algorithm so it makes sense.

Helper Box		
Hint	*Common Mistake*	*What to Look For*
Which step removes something you actually need?	Throwing away homework papers is the opposite of preparing a caddy.	A step that doesn't belong because it destroys or removes needed materials.

Activity A8 — Getting Ready for a Trip

Find the mistake in this algorithm.

1. Pack your suitcase.
2. Check the weather.
3. Load the suitcase into the car.
4. Zip the suitcase.

Rewrite the algorithm so it makes sense.

Helper Box		
Hint	*Common Mistake*	*What to Look For*
Which step should happen before packing?	Not noticing that checking the weather affects what you pack.	A step that belongs at the beginning.

Activity Set B: Fix the Algorithm

Activity B1 — Preparing for and Taking a Spelling Test

Fix the broken algorithm by rewriting it correctly.

1. Practice the words you don't know.
2. Look at the spelling list.
3. Ask someone to quiz you.

1. ____________________

2. ____________________

3. ____________________

4. ____________________

5. ____________________

Helper Box		
Hint	*Common Mistake*	*What to Look For*
What must happen before you can practice or be quizzed?	Starting with "take the test" instead of studying first.	Reviewing the list, then practicing, then quizzing, then taking the test.

Activity B2 — Feeding Backyard Chickens

Fix the broken algorithm by rewriting it correctly.

1. Scatter feed on the ground.
2. Fill the feeder with food.
3. Check the water container.
4. Refill the water container if it's empty.

1. ______________________________

2. ______________________________

3. ______________________________

4. ______________________________

5. ______________________________

Helper Box		
Hint	*Common Mistake*	*What to Look For*
Which steps belong together before feeding the chickens?	Feeding before checking water or filling the feeder.	A sequence that checks and fills water, fills the feeder, then feeds.

Activity B3 — Setting the Table

Fix the broken algorithm by rewriting it correctly.

1. Put plates on the table.
2. Put napkins on the table.
3. Put forks on the table.
4. Clear yesterday's dishes off the table.

1. ______________________________

2. ______________________________

3. ______________________________

4. ______________________________

5. ______________________________

Helper Box		
Hint	*Common Mistake*	*What to Look For*
What must happen before you can set the table?	Not noticing that the table must be cleared first.	A sequence that starts with clearing, then placing items.

Activity B4 — Building a LEGO Bridge

Fix the broken algorithm by rewriting it correctly.

1. Test the bridge by putting weight on it.
2. Snap bricks together randomly.
3. Gather the bricks you need.
4. Build the support pillars.

1. ______________________________

2. ______________________________

3. ______________________________

4. ______________________________

5. ______________________________

6. ______________________________

Helper Box		
Hint	*Common Mistake*	*What to Look For*
Which steps must happen before you can test the bridge?	Testing before building the structure properly.	Gathering materials, building supports, building the bridge, then testing it.

Activity B5 — Getting Ready for a Bike Ride

Fix the broken algorithm by rewriting it correctly.

1. Put on your helmet.
2. Ride your bike.
3. Check the tires.
4. Fill the water bottle.
5. Put on your shoes.

1. ______________________________
2. ______________________________
3. ______________________________
4. ______________________________
5. ______________________________

Helper Box		
Hint	*Common Mistake*	*What to Look For*
What must happen before you start riding?	Putting on the helmet too early or riding before checking the bike.	A sequence that includes checking equipment, getting ready, then riding.

Activity B6 — Making a Birdhouse

Fix the broken algorithm by rewriting it correctly.

1. Paint the birdhouse.
2. Hammer the pieces of wood together.
3. Gather the wood pieces and nails.
4. Let the paint dry.
5. Sand the edges of the wood.

1. ______________________________

2. ______________________________

3. ______________________________

4. ______________________________

5. ______________________________

6. ______________________________

Helper Box		
Hint	*Common Mistake*	*What to Look For*
What must happen before painting?	Painting before assembling or sanding.	A sequence with gathering → sanding → assembling → painting → drying.

Activity B7 — Preparing a Magic Show

Fix the broken algorithm by rewriting it correctly.

1. Perform the magic tricks.
2. Practice the tricks.
3. Set up the table and props.
4. Invite the audience to sit down.

1. ____________________
2. ____________________
3. ____________________
4. ____________________
5. ____________________

Helper Box		
Hint	*Common Mistake*	*What to Look For*
Which steps must happen before the audience arrives?	Performing before practicing or setting up.	A sequence that includes practice → setup → audience → performance.

Activity B8 — Preparing a Model Volcano

Fix the broken algorithm by rewriting it correctly.

1. Pour vinegar into the volcano.
2. Build the volcano shape with clay.
3. Add baking soda inside the volcano.
4. Place the volcano on a tray.
5. Mix food coloring into the vinegar.

1. ______________________________

2. ______________________________

3. ______________________________

4. ______________________________

5. ______________________________

6. ______________________________

Helper Box		
Hint	*Common Mistake*	*What to Look For*
What must happen before adding ingredients?	Pouring vinegar before the volcano is built or placed on a tray.	A sequence that includes building → placing → adding ingredients → eruption.

Activity Set C: Compare Expected vs. Actual

Activity C1 — Setting Up a Bird Feeder

You set up at bird feeder on the lamp post near the corner of a busy street.

Expected: Birds visit the feeder throughout the day.

Actual: No birds come at all.

What went wrong?

____

____

____

____

Helper Box		
Hint	*Common Mistake*	*What to Look For*
Think about where the feeder was placed.	Not noticing that birds are scared to use a feeder close to people or noise. birds away.	Identifying a placement or setup issue that prevents birds from visiting.

Activity C2 — Setting Up a Photography Corner

You cleared out a section in your closet to take photos.

Expected: Your photography corner has good lighting and a clean background.

Actual: Your photos look dark and cluttered.

What went wrong?

Helper Box		
Hint	*Common Mistake*	*What to Look For*
Which part of the setup affects how bright and clean the photos look?	A missing light source or messy background.	Identifying missing lighting or a cluttered background as the cause.

Activity C3 — Organizing a Puzzle Shelf

All the puzzles on your shelf fell down, the boxes came open, and pieces spilled everywhere. You picked up the pieces, put them in the boxes, and put the boxes back on the shelf.

Expected: All puzzles are sorted by size and stored neatly.

Actual: Puzzle pieces are still mixed together.

What went wrong?

____

____

____

____

Helper Box		
Hint	*Common Mistake*	*What to Look For*
Which step should happen before putting puzzles on the shelf?	Skipping the step where pieces are sorted into the correct boxes.	A clear explanation that sorting must happen before organizing.

Activity C4 — Setting Up a Science Notebook

You put all the sheets for your science notebook in your binder.

Expected: The notebook is ready for experiments.

Actual: The pages are out of order and the table of contents is wrong.

What went wrong?

Helper Box		
Hint	*Common Mistake*	*What to Look For*
Which steps must happen before writing the table of contents?	Not numbering pages or labeling sections first.	An explanation that recognizes the order of setup steps was incorrect.

Activity C5 — Setting Up a Mini Greenhouse

You built a greenhouse and put your plants inside.

Expected: The greenhouse stays warm and plants begin to sprout.

Actual: The plants dry out and the soil is cold.

What went wrong?

Helper Box		
Hint	*Common Mistake*	*What to Look For*
Think about what keeps a greenhouse warm and moist.	Not checking for a closed lid or placing the greenhouse in sunlight.	An explanation that identifies missing warmth or moisture steps.

Activity C6 — Preparing a Science Display Board

You glued the information on with the glue stick that was missing its lid.

Expected: The display board is neat and easy to read.

Actual: The information is crooked and some pieces fall off.

What went wrong?

Helper Box		
Hint	*Common Mistake*	*What to Look For*
Which step helps the pieces stay in place?	Skipping the step where the layout is planned before gluing.	Identifying poor layout planning or weak attachment.

Activity C7 — Setting Up a Marble Run

You built the marble run. One track piece is tilted or does not line up correctly.

Expected: The marble rolls smoothly from start to finish.

Actual: The marble gets stuck halfway through.

What went wrong?

____

____

____

____

Helper Box		
Hint	*Common Mistake*	*What to Look For*
Which part of the run controls how the marble moves?	Not noticing that one of the track pieces is tilted the wrong way.	An explanation that identifies a misaligned or uneven track section.

Activity C8 — Preparing a Pet Habitat

You put your new pet in a cardboard box under your bed.

Expected: The habitat is safe and comfortable for the pet.

Actual: The pet keeps escaping or knocking things over.

What went wrong?

Helper Box		
Hint	*Common Mistake*	*What to Look For*
Which step ensures the habitat is secure?	Not checking that the lid or door is fully closed or locked.	An explanation that identifies a missing safety or securing step.

Activity Set D: Debug a Full Scenario

Activity D1 — Preparing for a Class Presentation

This algorithm has **three mistakes**. Underline them, then rewrite the steps correctly.

1. Start presenting your project.
2. Gather your materials and notes.
3. Practice your presentation.
4. Create your slides.
5. Check that the computer and projector work.

1. ____

2. ____

3. ____

4. ____

5. ____

Helper Box		
Hint	*Common Mistake*	*What to Look For*
Which steps must happen before you begin presenting?	Not noticing that creating slides and practicing must happen before presenting.	Sequence including preparing, creating slides, practicing, checking, then presenting.

Activity D2 — Setting Up a Domino Chain Reaction

This algorithm has **two mistakes**. Underline them, then rewrite the steps correctly.

1. Knock over the first domino to test the chain.
2. Choose a flat surface.
3. Space the dominos evenly.
4. Gather the dominos you need.
5. Build the chain in a straight line.

1. ______________________________

2. ______________________________

3. ______________________________

4. ______________________________

5. ______________________________

Helper Box		
Hint	*Common Mistake*	*What to Look For*
Which steps must happen before testing the chain?	Testing before the chain is built or before choosing a surface.	Sequence beginning with gathering materials, preparing, then building, then testing.

Activity D3 — Preparing a Reading Nook

This algorithm has **three mistakes**. Underline them, then rewrite the steps correctly.

1. Put your books in a pile on the floor.
2. Turn on a lamp for good lighting.
3. Place pillows and blankets neatly.
4. Dump old papers onto the chair.
5. Choose a quiet spot to set up your nook.

1. ______________________________

2. ______________________________

3. ______________________________

4. ______________________________

5. ______________________________

Helper Box		
Hint	*Common Mistake*	*What to Look For*
Which steps make the space less comfortable instead of more?	Not noticing that dumping items or piling books on the floor creates clutter.	Choosing a spot → clearing clutter → arranging seating → adding lighting.

Activity D4 — Setting Up a Weather Observation Chart

This algorithm has **two mistakes**. Underline them, then rewrite the steps correctly.

1. Record today's weather on the chart.
2. Draw the chart grid on a poster.
3. Gather markers and a ruler.
4. Hang the chart on the wall.
5. Decide what weather symbols you will use.

1. ______________________________

2. ______________________________

3. ______________________________

4. ______________________________

5. ______________________________

Helper Box		
Hint	*Common Mistake*	*What to Look For*
Which steps must happen before you can record any weather?	Starting with recording data before the chart is even created.	Gathering materials → designing the chart → hanging it → recording.

Activity D5 — Setting Up a Mini Art Studio

This algorithm has **three mistakes**. Underline them, then rewrite the steps correctly.

1. Start painting your picture.
2. Lay down newspaper or a drop cloth.
3. Put brushes and paints in easy-to-reach spots.
4. Turn on a fan that blows directly at the wet paint.
5. Set up your easel and paper.

1. ____________________

2. ____________________

3. ____________________

4. ____________________

5. ____________________

Helper Box		
Hint	*Common Mistake*	*What to Look For*
Which steps should happen *before* you begin painting?	Not noticing that the fan blowing on wet paint is a problem.	Preparing workspace → arranging supplies → setting up easel → painting last.

Activity D6 — Preparing a Bird Migration Chart

This algorithm has **two mistakes**. Underline them, then rewrite the steps correctly.

1. Record bird sightings on the chart.
2. Draw the chart grid on a poster.
3. Research which birds migrate through your area.
4. Hang the blank poster on the wall.
5. Gather colored markers.

1. ______________________________

2. ______________________________

3. ______________________________

4. ______________________________

5. ______________________________

Helper Box		
Hint	*Common Mistake*	*What to Look For*
Which steps must happen before you can record any sightings?	Starting with recording data before the chart is created or before researching the birds.	Researching, gathering materials, creating chart, hanging it, recording sightings.

Activity D7 — Organizing a Science Supply Cabinet

This algorithm has **three mistakes**. Underline them, then rewrite the steps correctly.

1. Put random items on the top shelf.
2. Sort supplies into categories (safety gear, tools, materials).
3. Throw away expired or broken items.
4. Dump everything onto the floor.
5. Label each shelf clearly.

1. ______________________________

2. ______________________________

3. ______________________________

4. ______________________________

5. ______________________________

Helper Box		
Hint	*Common Mistake*	*What to Look For*
Which steps make the cabinet *less* organized instead of more?	Not noticing that dumping items or placing things randomly breaks organization.	Clearing clutter, sorting items, labeling shelves, and placing supplies neatly.

Activity D8 — Setting Up a Classroom Weather Station

This algorithm has **two mistakes**. Underline them, then rewrite the steps correctly.

1. Mount the thermometer and wind vane outside.
2. Record today's weather data.
3. Choose a safe outdoor location for the instruments.
4. Gather the tools you need to mount the equipment.
5. Build the weather station stand.

1. ______________________________

2. ______________________________

3. ______________________________

4. ______________________________

5. ______________________________

Helper Box		
Hint	*Common Mistake*	*What to Look For*
Which steps must happen before you can mount the instruments or record data?	Recording weather before the station is built or choosing the location too late.	Including gathering tools, choosing a spot, building the stand, and collecting data.

CERTIFICATE OF ACHIEVEMENT

Bug-Zapper Extraordinare

This certificate is awarded to ______________________________ for completing all activities in **Chapter 3: Debugging** of *Think Like a Coder*.

__________________________ Helper Signature

__________________________ Date

LOGIC LAB BOOKS

Chapter 4: Loops

Repeat with Reason

Introduction: Loops

Have you ever done something over and over again — like practicing a cartwheel, stacking blocks, or jumping rope? Maybe you've even said, "Ugh, I have to do this *again*?"

Congratulations — you've already experienced a loop.

Computers repeat things too, and they don't get tired, bored, or distracted by snacks. They repeat steps **exactly**, the same way, every time. And here's the catch: computers don't magically know *when* to repeat or *how many times*. You have to tell them.

That's where **loops** come in.

A loop is a way to repeat steps without writing them again and again. Instead of listing the same instructions five times, you can tell the computer:

"Repeat these steps 5 times."

Or:

"Repeat these steps until something happens."

Loops help computers:

- save time
- avoid mistakes
- follow patterns
- do big jobs with fewer instructions

Think about tying your shoes. You don't say, "Make a loop... wrap the lace... pull through..." every single time you practice. You just repeat the steps until you get it right.

Computers need that same idea — and written clearly.

In this chapter, you'll learn how loops work, how to spot them, and how to write your own. By the end, you'll be able to turn long, repetitive instructions into short, powerful loops.

Ready to make your code work smarter, not harder?

Let's go.

Mini-Lesson: What Is a Loop?

A **loop** is a set of steps that repeats.

That's it. Simple, and incredibly powerful.

There are two main kinds of loops you'll learn in this chapter:

1. Fixed Loops ("Repeat X Times")

These loops repeat a certain number of times.

Example:

Repeat 3 times:

- Clap your hands
- Stomp your feet

Instead of writing those steps three separate times, the loop does the repeating for you.

Computers love fixed loops because they're neat, tidy, and predictable.

2. "Repeat Until" Loops

These loops keep going **until something happens**.

Example:

Repeat until the cup is full:

- Pour a little water

The loop stops when the condition becomes true.

This kind of loop is great for:

- filling
- counting
- checking
- waiting
- watching for something to change

Why Loops Matter to Computers

Computers don't get bored. They don't lose count. They don't say, "Wait, did I already do this step?"

And they also don't know *when* to repeat unless you tell them.

Loops help computers:

- repeat steps without mistakes
- avoid super-long instructions
- follow patterns
- do big tasks efficiently

Without loops, even simple programs would be huge.

How Loops Help You Think Like a Programmer

When you learn loops, you learn how to:

- spot patterns
- simplify long instructions
- think in repetitions
- write cleaner, smarter code
- avoid repeating yourself

Programmers use loops constantly — in games, animations, robots, apps, and everything else.

Loops are one of the biggest "Aha!" moments in coding.

What You'll Do in This Chapter

You'll practice:

- finding loops hidden inside long instructions
- deciding when a loop is needed
- rewriting repeated steps using loops
- debugging loops that don't work correctly

By the end, you'll be able to turn long, messy instructions into short, powerful loops that computers love.

Ready to level up your coding superpowers? Let's jump in.

Activity Set A: Identify the Loop

Activity A1 — Jump Rope Practice

You're practicing jump rope. Read the steps below. Underline the steps that repeat.

1. Hold the jump rope handles
2. Swing the rope over your head
3. Jump once
4. Swing the rope over your head
5. Jump once
6. Swing the rope over your head
7. Jump once

Helper Box		
Hint	*Common Mistake*	*What to Look For*
Look for steps that happen again and again.	Kids sometimes underline all the steps.	The "swing rope → jump" pattern repeats.

Activity A2 — Watering the Garden Row

You're watering a row of plants. Underline the steps that repeat.

1. Fill the watering can
2. Pour water on Plant 1
3. Pour water on Plant 2
4. Pour water on Plant 3
5. Pour water on Plant 4
6. Put the watering can away

Helper Box		
Hint	*Common Mistake*	*What to Look For*
Which steps happen for each plant?	Kids may include "fill the watering can" — but that only happens once.	The "pour water" step repeats for each plant.

Activity A3 — Setting Up Dominoes

You're building a domino line. Underline the steps that repeat.

1. Pick up a domino
2. Place it standing on the floor
3. Pick up a domino
4. Place it standing on the floor
5. Pick up a domino
6. Place it standing on the floor

Helper Box		
Hint	*Common Mistake*	*What to Look For*
Look for pairs of steps that repeat together.	Underlining only one of the two steps.	"Pick up → place" repeats three times.

Activity A4 — Packing Party Favor Bags

You're helping pack party favor bags. Underline the steps that repeat.

1. Open an empty bag
2. Put in a sticker sheet
3. Put in a small toy
4. Put in a piece of candy
5. Tie the bag closed
6. Open the next empty bag
7. Put in a sticker sheet
8. Put in a small toy
9. Put in a piece of candy
10. Tie the bag closed

Helper Box		
Hint	*Common Mistake*	*What to Look For*
Look for a group of steps that repeat in the same order.	Steps 1–5 repeat as 6–10.	Kids may forget that "open bag" and "tie bag" are part of the loop.

Activity A5 — Practicing Piano Notes

You're practicing a short pattern on the piano. Underline the steps that repeat.

1. Play Middle C
2. Play D
3. Play E
4. Play D
5. Play Middle C
6. Play D
7. Play E
8. Play D
9. Play Middle C

Helper Box		
Hint	*Common Mistake*	*What to Look For*
Look for a musical pattern that repeats.	Circling only the middle part of the pattern.	C → D → E → D repeats twice. The final Middle C comes after the repeated pattern.

Activity A6 — Feeding the Classroom Fish

You're helping feed the fish in the classroom tank. Underline the steps that repeat.

1. Open the container of fish food
2. Pinch a small amount of food
3. Sprinkle it into the tank
4. Watch the fish eat
5. Pinch a small amount of food
6. Sprinkle it into the tank
7. Watch the fish eat
8. Close the container of fish food

Helper Box		
Hint	*Common Mistake*	*What to Look For*
Look for the repeated "pinch → sprinkle → watch" cycle.	Including "open" or "close" in the loop — those happen once.	Steps 2–4 repeat as 5–7.

Activity A7 — Building a Block Tower

You're building a tall block tower. Underline the steps that repeat.

1. Pick up a block
2. Place it on the tower
3. Check if the tower is leaning
4. Pick up a block
5. Place it on the tower
6. Check if the tower is leaning
7. Pick up a block
8. Place it on the tower
9. Check if the tower is leaning
10. Blow gently to test stability

Helper Box		
Hint	*Common Mistake*	*What to Look For*
Look for a repeating *three step* pattern.	Kids may think "check if leaning" is separate even though it repeats.	Steps 1–3 repeat twice more.

Activity A8 — Sorting a Deck of Cards

You're sorting cards into piles by color. <u>Underline</u> the steps that repeat.

1. Pick up the top card
2. Look at its color
3. If it's red, place it in the red pile
4. If it's black, place it in the black pile
5. Pick up the next card
6. Look at its color
7. If it's red, place it in the red pile
8. If it's black, place it in the black pile
9. Pick up the next card
10. Look at its color
11. If it's red, place it in the red pile
12. If it's black, place it in the black pile

Helper Box		
Hint	*Common Mistake*	*What to Look For*
Look for the repeated "pick up → look → sort" cycle.	Kids may underline only the "sort" steps — but the whole cycle repeats.	Steps 1–4 repeat twice more.

Activity Set B: Expand the Loop

Example

Here is a loop. Write out the steps.

Repeat 4 times.

- Tap your desk once
- Tap your nose twice

Tap your desk once. Tap your nose twice. Tap your desk once. Tap your nose twice.

Tap your desk once. Tap your nose twice. Tap your desk once. Tap your nose twice.

Activity B1 — Jumping Jacks Warm-Up

Here is a loop. Write out the steps.

Repeat 3 times:

- Do one jumping jack

Helper Box		
Hint	*Common Mistake*	*What to Look For*
Write "Do one jumping jack" three times.	Kids sometimes write "Do 3 jumping jacks" instead of expanding the loop.	Three separate sentences, each saying the same step.

Activity B2 — Feeding the Hamster

Here is a loop. Write out the steps.

Repeat 2 times:

- Scoop a small amount of food
- Pour it into the bowl

Helper Box		
Hint	*Common Mistake*	*What to Look For*
Each cycle has *two* steps.	Forgetting to repeat both steps.	Two full cycles of "scoop → pour."

Activity B3 — Practicing a Dance Move

Here is a loop. Write out the steps.

Repeat 4 times:

- Step to the left
- Clap once

Helper Box		
Hint	*Common Mistake*	*What to Look For*
You should have *eight* total steps.	Kids sometimes write only the first step four times.	"Step → clap" repeated four times.

Activity B4 — Decorating Cupcakes

Here is a loop. Write out the steps.

Repeat 3 times:

- Pick up a cupcake
- Add frosting
- Add sprinkles on top

Helper Box		
Hint	*Common Mistake*	*What to Look For*
This loop has *three* steps per cycle.	Kids may forget the "pick up a cupcake" step in later cycles.	Three full cycles of "pick up → frost → sprinkle."

Activity B5 — Cleaning Up the Playroom

Here is a loop. Write out the steps.

Repeat 4 times:

- Pick up a toy
- Decide which bin it belongs in
- Place it in the correct bin

Helper Box		
Hint	*Common Mistake*	*What to Look For*
Each toy follows the same three-step sorting routine.	Kids may skip the step where they decide which bin to use.	Four complete pick up → choose bin → put away cycles.

Activity B6 — Practicing a Drum Pattern

Here is a loop. Write out the steps.

Repeat 2 times:

- Hit the snare drum
- Hit the snare drum
- Hit the cymbal
- Pause for one beat

Helper Box		
Hint	*Common Mistake*	*What to Look For*
Notice the snare drum happens twice in each cycle.	Kids may combine the two snare hits into one step.	Two full cycles of four steps each.

Activity B7 — Checking Windows Before Leaving

Here is a loop. Write out the steps.

Repeat 3 times:

- Walk to the next window
- Check whether it is closed.
- Close it if needed.
- Check whether it is locked.
- Lock it if needed.

Helper Box		
Hint	*Common Mistake*	*What to Look For*
Repeat all five steps for each window.	Skipping a check or assuming every window needs the same action.	Fifteen total steps (5 steps for each of the 3 windows).

Activity B8 — Sorting Puzzle Pieces

Here is a loop. Write out the steps.

Repeat 2 times:

- Pick up a puzzle piece
- Look at its color
- If it's blue, place it in the blue pile
- If it's green, place it in the green pile
- If it's red, place it in the red pile

Helper Box		
Hint	*Common Mistake*	*What to Look For*
There should be 10 steps.	Repeating only one color check per loop.	Two full cycles of five steps each.

Activity Set C: Rewrite Using Loops

Example

Rewrite the steps using a loop.

- Tap your desk once
- Tap your desk once

repeat 2 times

Tap your desk once

end loop

Activity C1 — Bouncing a Ball

Rewrite the steps using a loop.

1. Bounce the ball once
2. Bounce the ball once
3. Bounce the ball once

Helper Box		
Hint	*Common Mistake*	*What to Look For*
Count how many times the same step appears.	Kids may write "Bounce the ball 3 times" without using loop wording.	"Repeat 3 times: bounce the ball once."

Activity C2 — Brushing the Dog

Rewrite the steps using a loop.

1. Brush the dog's fur
2. Brush the dog's fur
3. Brush the dog's fur
4. Brush the dog's fur

Helper Box		
Hint	*Common Mistake*	*What to Look For*
Look for identical steps.	Forgetting to include the step inside the loop.	"Repeat 4 times: brush the dog's fur."

Activity C3 — Practicing a Cheer

Rewrite the steps using a loop.

1. Say "Go team!"
2. Say "Go team!"
3. Say "Go team!"

Helper Box		
Hint	*Common Mistake*	*What to Look For*
Count the number of repeats.	Kids may add extra words or change the step.	"Repeat 3 times: say 'Go team!'"

Activity C4 — Testing Robot Movements

You need to test your robot's movement pattern. Rewrite the steps using a loop.

1. Move forward 3 steps
2. Turn right
3. Flash your headlight
4. Move forward 3 steps
5. Turn right
6. Flash your headlight
7. Move forward 3 steps
8. Turn right
9. Flash your headlight

Helper Box		
Hint	*Common Mistake*	*What to Look For*
Look for a repeating *group* of steps.	Kids may only loop the "move forward" step.	Repeat 3 times: move forward → turn → headlight.

Activity C5 — Painting Fence Boards

Rewrite the steps using a loop.

1. Dip the brush in paint
2. Paint one board
3. Dip the brush in paint
4. Paint one board
5. Dip the brush in paint
6. Paint one board
7. Dip the brush in paint
8. Paint one board

____

____

____

Helper Box		
Hint	*Common Mistake*	*What to Look For*
Notice the two step pattern.	Forgetting to include both steps inside the loop.	Repeat 4 times: dip → paint."

Activity C6 — Sharpening a Pencil

Rewrite the steps using a "repeat until" loop.

1. Turn the pencil in the sharpener
2. Check the tip
3. Turn the pencil in the sharpener
4. Check the tip
5. Turn the pencil in the sharpener
6. Check the tip

____

____

____

Helper Box		
Hint	*Common Mistake*	*What to Look For*
This loop stops when the pencil is sharp.	Kids may try to count the repeats instead of using a condition.	"Repeat until the pencil is sharp: turn → check."

Activity C7 — Checking Classroom Plants

Rewrite the steps using a loop.

1. Walk to Plant 1
2. Touch the soil
3. If it's dry, add water
4. Walk to Plant 2
5. Touch the soil
6. If it's dry, add water
7. Walk to Plant 3
8. Touch the soil
9. If it's dry, add water

Helper Box		
Hint	*Common Mistake*	*What to Look For*
Even though there's an "if," the pattern still repeats.	Kids may think conditions can't be inside loops.	"Repeat 3 times: walk → touch → if dry, water."

Activity C8 — Sorting Books on a Shelf

Rewrite the steps using a loop.

1. Pick up a book
2. Look at the first letter of the title
3. If it starts with A–M, place it on the left side
4. If it starts with N–Z, place it on the right side
5. Pick up a book
6. Look at the first letter of the title
7. If it starts with A–M, place it on the left side
8. If it starts with N–Z, place it on the right side
9. Pick up a book
10. Look at the first letter of the title
11. If it starts with A–M, place it on the left side
12. If it starts with N–Z, place it on the right side

Helper Box		
Hint	*Common Mistake*	*What to Look For*
Even though the "if" steps branch, they are still part of the loop.	Kids may write only one alphabetical option.	"Repeat 3 times: pick up → look → if A-M, left →if N-Z, right."

Activity Set D: Debug the Loop

Activity D1 — Feeding the Cat

Fix the loop so it works correctly.

Broken Loop:

Repeat 3 times:

- Scoop food
- Scoop food
- Pour into bowl

What's wrong? One step is repeated when it shouldn't be.

Rewrite the loop so it works correctly.

Helper Box		
Hint	*Common Mistake*	*What to Look For*
Each cycle should have "scoop → pour."	Kids may keep both "scoop" steps.	Only one "scoop" per cycle.

Activity D2 — Practicing a Wave Motion

Fix the loop so it works correctly.

Broken Loop:

Repeat 4 times:

- Wave your hand
- Wave your hand
- Wave your hand

What's wrong? The loop repeats too many times inside itself.

Rewrite the loop so it works correctly.

Helper Box		
Hint	*Common Mistake*	*What to Look For*
The loop says "repeat 4 times," so the inside should be one wave.	Leaving all three waves inside the loop.	"Repeat 4 times: wave your hand."

Activity D3 — Watering the Plants

Fix the loop so it works correctly.

Broken Loop:

Repeat 3 times:

- Move to the next plant
- Water the plant
- Check whether another plant remains

What's wrong? There are only two plants, so the loop repeats too many times.

Rewrite the loop so it works correctly.

Helper Box		
Hint	*Common Mistake*	*What to Look For*
Match the repetition count to the number of plants.	Keeping the repetition count at 3.	A loop that repeats once for each of the two plants.

Activity D4 — Cleaning the Whiteboard

Fix the loop so it works correctly.

Broken Loop:

Repeat 3 times:

- Move to the next section
- Wipe the section
- Check that the section is clean
- Wipe the section
- Wipe the section

What's wrong? The loop wipes each section two extra times.

Rewrite the loop so it works correctly.

Helper Box		
Hint	*Common Mistake*	*What to Look For*
Each repetition should clean one section.	Leaving either extra "Wipe the section" step in the loop.	A three-step move → wipe → check cycle.

Activity D5 — Practicing a Drum Rhythm

Fix the loop so it works correctly.

Broken Loop:

Repeat 2 times:

- Hit the snare
- Hit the cymbal
- Hit the snare
- Hit the cymbal

What's wrong? The pattern is doubled inside the loop.

Rewrite the loop so it works correctly.

Helper Box		
Hint	*Common Mistake*	*What to Look For*
Look for the intended pattern: snare → cymbal.	Keeping both snares and both cymbals.	One snare and one cymbal per cycle.

Activity D6 — Sorting Stickers

Fix the loop so it works correctly.

Broken Loop:

Repeat 5 times:

- Pick up a sticker
- If it's shiny, put it in the shiny pile
- If it's not shiny, put it in the regular pile
- Pick up a sticker

What's wrong? "Pick up a sticker" appears twice.

Rewrite the loop so it works correctly.

Helper Box		
Hint	*Common Mistake*	*What to Look For*
Each cycle should start with one "pick up."	Leaving both "pick up" steps.	Remove the extra "pick up."

Activity D7 — Building a LEGO Wall

Fix the loop so it works correctly.

Broken Loop:

Repeat 4 times:

- Pick up a brick
- Place it on the wall
- Check if the wall is straight
- Pick up a brick
- Place it on the wall

What's wrong? The loop repeats part of the pattern inside itself.

Rewrite the loop so that it works correctly.

Helper Box		
Hint	*Common Mistake*	*What to Look For*
Look for the intended cycle: pick up → place → check.	Keeping the extra "pick up → place."	Only one full cycle per loop repeat.

Activity D8 — Checking the Weather Sensors

Fix the loop so it works correctly.

Broken Loop:

Repeat 3 times:

- Walk to Sensor A
- Read the temperature
- Walk to Sensor B
- Read the temperature
- Walk to Sensor A

What's wrong? The loop returns to Sensor A twice.

Rewrite the loop so it works correctly.

Helper Box		
Hint	*Common Mistake*	*What to Look For*
Each cycle should check both sensors once.	Leaving the extra "Walk to Sensor A."	A clean cycle: A → read → B → read.

Activity D9 — Filling a Bird Feeder

Fix the loop so it works correctly.

Broken Loop:

Repeat 4 times:

- Scoop birdseed
- Pour it into the feeder
- Check if the feeder is full

What's wrong? This shouldn't repeat a fixed number of times — it should repeat *until the feeder is full*.

Rewrite the loop so it works correctly.

____

____

____

____

____

____

Helper Box		
Hint	*Common Mistake*	*What to Look For*
The loop should stop when the feeder is full, not after 4 scoops.	Keeping "Repeat 4 times" instead of using a condition.	A "repeat until full" loop.

Activity Set E: Write Your Own Loops

Activity E1 — Morning Stretching

Write a loop for this stretching routine:

- Touch your toes
- Stand up straight

Helper Box		
Hint	*Common Mistake*	*What to Look For*
Decide how many times the stretch repeats.	Kids may list the steps instead of using loop wording.	"Repeat X times: touch → stand."

Activity E2 — Feeding the Ducks

Write a loop for feeding ducks at the pond.

Helper Box		
Hint	*Common Mistake*	*What to Look For*
Think about what you do again and again.	Writing a whole story instead of steps.	A simple repeating action like "throw food."

Activity E3 — Practicing a Magic Trick

Write a loop using these steps:

- Wave your wand
- Say "Abracadabra!"

Helper Box		
Hint	*Common Mistake*	*What to Look For*
Choose how many times the trick repeats.	Forgetting to include both steps inside the loop.	"Repeat X times: wave → say."

Activity E4 — Cleaning Your Desk

Write a loop for cleaning your desk. Choose steps that repeat.

Helper Box		
Hint	*Common Mistake*	*What to Look For*
Think about actions you do for each item.	Kids may include "throw away trash" even if it only happens once.	A repeating cycle like "pick up → sort → place."

Activity E5 — Building a Snowman

Write a loop using these steps:

- Roll a snowball
- Pack it tighter
- Check if it's big enough

Helper Box		
Hint	*Common Mistake*	*What to Look For*
The loop should stop when the snowball reaches the right size.	Forgetting to include the "check" step.	A "repeat until" loop.

Activity E6 — Practicing a Piano Pattern

Write a loop for practicing a short musical pattern of your choice.

Helper Box		
Hint	*Common Mistake*	*What to Look For*
Choose 2–3 notes that repeat.	Writing too many different notes.	A clear, repeating musical cycle.

Activity E7 — Checking Robot Sensors

Write a loop using these steps:

- Read the left sensor
- Read the right sensor
- Compare the readings

Helper Box		
Hint	*Common Mistake*	*What to Look For*
This loop repeats until the readings match.	Kids may forget to use "repeat until."	A loop that stops when a condition is met.

Activity E8 — Sorting Recyclables

Write a loop for sorting items into recycling bins. Choose the steps and decide when the loop stops.

Helper Box		
Hint	*Common Mistake*	*What to Look For*
Think about what repeats for *each* item.	Kids may forget to include a stopping condition.	A cycle like "pick up → check → sort," with a clear end.

CERTIFICATE OF ACHIEVEMENT

Loop Guru

This certificate is awarded to ______________________________ for completing all activities in **Chapter 4: Loops** of *Think Like a Coder*.

___________________________ Helper Signature

___________________________ Date

LOGIC LAB BOOKS

Chapter 5: Conditions

Code that Thinks

Introduction: Conditions

Have you ever made a choice because something *happened*?

- **If** it's raining, you grab an umbrella.
- **If** you're hungry, you get a snack.
- **If** the light is red, you stop.
- **If** your friend waves at you, you wave back.

You make decisions like this all day long — and so do computers.

Computers don't just follow steps in order. They don't just repeat patterns. Sometimes, they need to **check something** and then decide what to do next.

That's where **conditions** come in.

A condition is a rule that says:

"If this is true, do this."

It's like giving your computer a tiny brain that can notice things and react.

In this chapter, you'll learn how to:

- read conditions
- understand what a computer is checking
- follow both sides of a decision
- fix broken conditions
- and write your own IF/ELSE rules

By the end, you'll be able to teach a computer how to make choices — just like you do every day.

Mini-Lesson: How Conditions Work

Computers are great at following steps, and they're even better at making decisions when you give them clear rules.

A **condition** is a special instruction that checks something and then chooses what to do.

There are two main kinds:

1. `if` Statements

An `if` statement checks one thing.

If the door is closed, knock.

The computer looks at the condition ("Is the door closed?"). If it's true, it follows the step. If it's not true, it simply skips it.

This is like saying:

- "If it's cold, put on a jacket."
- "If the timer rings, take the cookies out."
- "If the robot sees an obstacle, stop."

Only one path. Only one action. Only when the condition is true.

2. `if/else` Statements

Sometimes you want the computer to choose between **two** actions.

That's where `if/else` comes in.

If the light is green, go. Else, stop.

The computer checks the condition:

- If it's true → do the first thing
- If it's false → do the second thing

This is how computers make decisions every second.

More examples:

- **If** the plant is dry, water it. **Else**, leave it alone.
- **If** the password is correct, log in. **Else**, try again.
- **If** the robot sees a wall, turn left. **Else**, keep going straight.

`if/else` gives your computer two possible paths — like a fork in the road.

What is a Condition?

A **condition** is the thing the computer checks.

It's always something that can be **true** or **false**.

- Is the cup full?
- Is the button pressed?
- Is the number bigger than 10?
- Is the door open?
- Is the robot touching the line?

Computers love true/false questions. They're simple, clear, and easy to check.

Why Conditions Matter

Conditions let computers:

- react
- choose
- respond
- behave differently in different situations

Without conditions, a computer would do the same thing every time — no matter what.

With conditions, your programs become smarter, more flexible, and more fun.

Your Turn

In the activities ahead, you'll:

- spot the condition
- expand both sides of a decision
- rewrite instructions using `if` and `if/else`
- debug broken conditions
- and write your own rules from scratch

You're about to teach your computer how to *think*.

Activity Set A: Identify the Condition

Activity A1 — Robot Battery Check

Underline the condition.

If the battery is low, charge the robot.

Helper Box		
Hint	*Common Mistake*	*What to Look For*
The condition is the part that follows "if."	Underlining the action instead of the condition.	The battery is low.

Activity A2 — Crossing the Street

Underline the condition.

If the crossing guard signals that it is safe, cross the street.

Helper Box		
Hint	*Common Mistake*	*What to Look For*
What must happen before it is safe to cross?	Underlining "cross the street."	The crossing guard signals that it is safe.

Activity A3 — Watering a Plant

Underline the condition.

If the soil is dry, water the plant.

Helper Box		
Hint	*Common Mistake*	*What to Look For*
Find the part that can be true or false.	Underlining the complete sentence.	The soil is dry.

Activity A4 — Robot Movement

Underline the condition.

If the robot sees a wall, turn left.

Helper Box		
Hint	*Common Mistake*	*What to Look For*
Look for what the robot checks before it turns.	Underlining only "a wall."	The robot sees a wall.

Activity A5 — Finishing a Painting

<u>Underline</u> the condition.

If the paint is dry, add the stickers.

Helper Box		
Hint	*Common Mistake*	*What to Look For*
What must be true before the stickers are added?	Underlining "add the stickers."	<u>The paint is dry.</u>

Activity A6 — Sorting Toys

<u>Underline</u> the condition.

If the toy is blue, put it in the blue bin.

Helper Box		
Hint	*Common Mistake*	*What to Look For*
Look for the test that controls where the toy goes.	Underlining the action instead of the condition.	<u>The toy is blue.</u>

Activity A7 — Password Check

Underline the condition.

If the password is correct, log in.

Helper Box		
Hint	*Common Mistake*	*What to Look For*
What must the computer verify?	Underlining "log in."	The password is correct.

Activity A8 — Weather Decision

Underline the condition.

Open the closet. If it is raining, take an umbrella. Close the closet.

Helper Box		
Hint	*Common Mistake*	*What to Look For*
Find the part that follows "if."	Underlining the entire sequence or the umbrella action.	It is raining.

Activity Set B: Trace the Decision

Complete both sides of each decision. Write what makes the condition true or false, then write what happens.

Example

If the library is open, return your book.

True	False
Condition: The library is open.	Condition: The library is closed.
Action: Return your book.	Action: Do not return your book.

Activity B1 — Robot Battery Check

Complete both sides. Write the condition state, then the action.

If the battery is low, charge the robot.

True	False

Helper Box		
Hint	*Common Mistake*	*What to Look For*
Describe the battery state before writing the action.	Writing only the action in each column.	Low → charge; not low → no charge.

Activity B2 — Crossing the Street

Complete both sides. Write the condition state, then the action.

If the crossing guard signals that it is safe, cross the street.

True	False
______________________	______________________
______________________	______________________

Helper Box		
Hint	*Common Mistake*	*What to Look For*
The crossing guard's signal determines the result.	Writing "safe" without identifying the signal.	Signal → cross; no signal → do not cross.

Activity B3 — Watering a Plant

Complete both sides. Write the condition state, then the action.

If the soil is dry, water the plant.

True	False
______________________	______________________
______________________	______________________

Helper Box		
Hint	*Common Mistake*	*What to Look For*
Describe the soil on each side.	Watering the plant for both results.	Dry → water; not dry → no water.

Activity B4 — Robot Movement

Complete both sides. Write the condition state, then the action.

If the robot sees a wall, turn left.

True	**False**
____________________	____________________
____________________	____________________

Helper Box		
Hint	*Common Mistake*	*What to Look For*
A false condition skips the conditional action.	Assuming the code says to move straight.	Wall → turn left; no wall → skip the turn.

Activity B5 — Finishing a Painting

Complete both sides. Write the condition state, then the action.

If the paint is dry, add the stickers.

True	**False**
____________________	____________________
____________________	____________________

Helper Box		
Hint	*Common Mistake*	*What to Look For*
Decide whether the stickers are added now or later.	Adding stickers while the paint is wet.	Dry → add; not dry → wait.

Activity B6 — Sorting Toys

Complete both sides. Write the condition state, then the action.

If the toy is blue, put it in the blue bin. Else, put it in the non-blue bin.

True	False
________________	________________
________________	________________

Helper Box		
Hint	*Common Mistake*	*What to Look For*
\`else\` supplies the false action.	Treating "not blue" as one specific color.	Blue → blue bin; otherwise → non-blue bin.

Activity B7 — Password Check

Complete both sides. Write the condition state, then the action.

If the password is correct, log in. Else, show an error message.

True	False
________________	________________
________________	________________

Helper Box		
Hint	*Common Mistake*	*What to Look For*
Each result has a different action.	Writing "log in" for both results.	Correct → log in; incorrect → error.

Activity B8 — Weather Decision

Complete both sides. Write the condition state, then the action.

If it is raining, choose the indoor activity. Else, choose the playground activity.

True	**False**
____________________	____________________
____________________	____________________

Helper Box		
Hint	*Common Mistake*	*What to Look For*
Match each activity with the weather condition.	Reversing the two activities.	Rain → indoors; otherwise → playground.

Activity Set C: Build the Decision

Rewrite each set of instructions using `if` or `if/else`.

Example

Check whether the supply bin is empty.

- If it is empty, refill it.
- If it is not empty, continue working.

```
if the supply bin is empty
    refill the supply bin
else
    continue working
end if
```

Activity C1 — Robot Battery Check

Check the robot's battery.

- If the battery is low, charge the robot.

Rewrite the instructions using `if`.

Helper Box		
Hint	*Common Mistake*	*What to Look For*
Place the condition after `if`.	Writing the action as the condition.	Battery low → charge.

Activity C2 — Watering a Plant

Check the soil.

- If the soil is dry, water the plant.

Rewrite the instructions using `if`.

Helper Box		
Hint	*Common Mistake*	*What to Look For*
The action belongs beneath the condition.	Leaving out `end if`.	Soil dry → water.

Activity C3 — Robot Movement

Check whether the robot sees a wall.

- If the robot sees a wall, turn left.

Rewrite the instructions using `if`.

Helper Box		
Hint	*Common Mistake*	*What to Look For*
Write the check, then the action.	Placing "turn left" after `if`.	Wall detected → turn left.-

Activity C4 — Returning a Library Book

Check the book's due date.

- If the book is due today, return it.

Rewrite the instructions using `if`.

Helper Box		
Hint	*Common Mistake*	*What to Look For*
The due date controls the action.	Adding an action the instructions do not provide.	Due today → return book.

Activity C5 — Sorting Shape Cards

Check the shape on each card.

- If the card shows a circle, place it in the circle pile.
- Otherwise, place it in the other-shapes pile.

Rewrite the instructions using `if/else`.

Helper Box		
Hint	*Common Mistake*	*What to Look For*
"Otherwise" becomes `else`.	Putting both actions in one branch.	Circle → circle pile; otherwise → other pile.

Activity C6 — Scanning a Ticket

Scan the ticket.

- If the ticket is valid, open the gate.
- Otherwise, show an error message.

Rewrite the instructions using `if/else`.

Helper Box		
Hint	*Common Mistake*	*What to Look For*
Give each result its own action.	Reversing the two actions.	Valid → open; otherwise → error.

Activity C7 — Choosing a Trail

Check whether the creek trail is open.

- If it is open, use the creek trail.
- Otherwise, use the meadow trail.

Rewrite the instructions using `if/else`.

Helper Box		
Hint	*Common Mistake*	*What to Look For*
Use the trail's status as the condition.	Writing a second `if` instead of `else`.	Open → creek; otherwise → meadow.-

Activity C8 — Controlling a Greenhouse Vent

Check the greenhouse temperature.

- If the temperature is above 80 degrees, open the vent.
- Otherwise, keep the vent closed.

Rewrite the instructions using `if/else`.

Helper Box		
Hint	*Common Mistake*	*What to Look For*
Compare the temperature with 80 degrees.	Reversing the vent actions.	Above 80 → open; otherwise → closed.

Activity Set D: Debug the Decision

Activity D1 — Wrong Condition

The robot should charge when its battery is low.

```
if the battery is full
    charge the robot
end if
```

What's wrong? The condition checks the wrong battery state.

Rewrite the code.

____

____

____

____

Helper Box		
Hint	*Common Mistake*	*What to Look For*
Compare the intended result with the condition.	Changing the action instead of the condition.	Change `full` to `low`.

Activity D2 — Missing `else`

A ticket scanner should open the gate for a valid ticket and show an error message for an invalid ticket.

```
if the ticket is valid
    open the gate
end if
```

What's wrong? One possible result is missing.

Rewrite the code.

Helper Box		
Hint	*Common Mistake*	*What to Look For*
Add a second path for an invalid ticket.	Putting the error inside the `if` branch.	Valid → open; `else` → error.

Activity D3 — Actions in the Wrong Branches

You may enter the library when it is open. You should wait outside when it is closed.

```
if the library is open
    wait outside
else
    enter the library
end if
```

Find the mistake and rewrite the code.

____

____

____

Helper Box		
Hint	*Common Mistake*	*What to Look For*
Check which action belongs with each result.	Changing the condition instead of the actions.	Open → enter; `else` → wait.

Activity D4 — Both Branches Do the Same Thing

Circle cards belong in the circle pile. All other cards belong in the other-shapes pile.

```
if the card shows a circle
    place it in the circle pile
else
    place it in the circle pile
end if
```

Find the mistake and rewrite the code.

Helper Box		
Hint	*Common Mistake*	*What to Look For*
Each branch needs a different destination.	Leaving the same action in both branches.	`else` → other-shapes pile.

Activity D5 — Actions Outside the Condition

The robot should stop only when it sees a wall. Otherwise, it should continue moving.

```
if the robot sees a wall
end if
stop the robot
continue moving
```

Find the mistake and rewrite the code.

Helper Box		
Hint	*Common Mistake*	*What to Look For*
Place each action inside its branch.	Moving only one action.	Wall → stop; `else` → continue.

Activity D6 — Condition Is Too Vague

You are choosing a snack for a table where no nuts are allowed.

```
if the snack is good
    place it on the table
else
    choose another snack
end if
```

Replace the vague condition with one a computer could check clearly.

Helper Box		
Hint	*Common Mistake*	*What to Look For*
Name a property that can be checked.	Replacing "good" with another opinion.	Use a clear nut-free condition.

Activity D7 — Condition Checks an Unrelated Detail

You should take a raincoat when it is raining.

```
if the raincoat is blue
    take the raincoat
else
    leave the raincoat
end if
```

Find the mistake and rewrite the code.

Helper Box		
Hint	*Common Mistake*	*What to Look For*
The condition must control the intended decision.	Changing the actions.	Check whether it is raining.

Activity D8 — Two Mistakes

The greenhouse vent should open when the temperature is above 80 degrees. Otherwise, it should remain closed.

```
if the temperature is below 80 degrees
    open the vent
end if
keep the vent closed
```

Find both mistakes and rewrite the code.

Helper Box		
Hint	*Common Mistake*	*What to Look For*
Check the comparison and second action.	Correcting only one mistake.	Above 80 → open; `else` → closed.

Activity Set E: Create Your Own Conditions

Activity E1 — Write an `if` Statement

Write an `if` statement that tells you when to pack a library book in your backpack.

```
if
________________________________________

________________________________________
end if
________________________________________
```

Helper Box		
Hint	*Common Mistake*	*What to Look For*
Think of something that can be true or false.	Writing an action after `if`.	One clear condition and action.

Activity E2 — Choose a Clear Condition

A fish should be fed at its scheduled feeding time.

Complete the condition.

```
if ____________________
    feed the fish
end if
```

Helper Box		
Hint	*Common Mistake*	*What to Look For*
Use the fish's feeding schedule.	Using an unrelated detail about the tank.	A scheduled-time condition.

Activity E3 — Write an `if/else` Statement

Write an `if/else` statement that sorts a card into one of two piles:

- A card with a star goes in the star pile.
- Any other card goes in the other-shapes pile.

```
if
______________________________

______________________________
else
______________________________

______________________________
end if
______________________________
```

Helper Box		
Hint	*Common Mistake*	*What to Look For*
Use the shape as the condition.	Putting both actions in one branch.	Star → star pile; `else` → other pile.

Activity E4 — Write a Matching Situation

Read the code.

```
if the classroom door is open
    enter the classroom
else
    wait in the hallway
end if
```

Write one situation in which `wait in the hallway` will run.

Helper Box		
Hint	*Common Mistake*	*What to Look For*
The `else` action runs when the condition is false.	Writing that the door is open.	The classroom door is closed.

Activity E5 — Make the Condition Clear

This condition is too vague:

```
if the backpack is ready
    leave for school
end if
```

Replace `the backpack is ready` with a condition that a computer could check clearly.

Helper Box		
Hint	*Common Mistake*	*What to Look For*
Name something observable about the backpack.	Replacing "ready" with another opinion.	A specific, testable condition.

Activity E6 — Use a Number in a Condition

Write an `if/else` statement that:

- Turns on a fan when the greenhouse temperature is above 80 degrees.
- Turns off the fan otherwise.

Helper Box		
Hint	*Common Mistake*	*What to Look For*
Compare the temperature with 80 degrees.	Reversing the fan actions.	Above 80 → on; `else` → off.

Activity E7 — Create Two Different Actions

Choose one topic:

- Selecting a game
- Organizing art supplies
- Choosing indoor or outdoor recess
- Getting ready for bed

Write an `if/else` statement with one clear condition and two different actions.

Helper Box		
Hint	*Common Mistake*	*What to Look For*
Decide what happens for both results.	Writing the same action in both branches.	One condition and two logical actions.

Activity E8 — Create Your Own Decision

Create a short conditional algorithm about a decision you make at home, at school, or during an activity.

Your algorithm must include:

- One clear condition
- An action for when the condition is true
- A different action for when the condition is false
- Complete `if/else` structure

Explain what makes your condition true or false.

Helper Box		
Hint	*Common Mistake*	*What to Look For*
Make the condition easy to evaluate.	Explaining the topic instead of the condition.	Code and explanation agree.

CERTIFICATE OF ACHIEVEMENT

If-Then Innovator

This certificate is awarded to ______________________________ for completing all activities in **Chapter 5: Conditions** of *Think Like a Coder*.

___________________________ Helper Signature

___________________________ Date

LOGIC LAB BOOKS

Chapter 6: Variables

Where Ideas Live

Introduction: Variables

Have you ever needed to *keep track* of something?

- How many points you've scored in a game
- How much money is in your piggy bank
- How many pages you've read
- How many cookies are left in the box
- How old you are

All of these things are pieces of information that can **change**.

You store them in your brain, or in a notebook, or on a scoreboard.

Computers need a way to store changing information too. They can't use pockets or notebooks, so they use something called a **variable**.

A variable is like a little container inside the computer. Just like a real container needs a label so you know what's inside, a variable needs a good name. If you name a variable `score`, you'll remember it holds points. If you name it `cookies`, you'll remember it counts cookies. A clear name helps both you *and the computer* understand what the variable is tracking. It holds a piece of information — and that information can change over time.

☒

In this chapter, you'll learn how to:

- understand what a variable is
- follow how a variable changes
- use variables to make decisions
- fix broken variable logic
- write your own instructions using variables

By the end, you'll be able to teach a computer how to **remember things**.

Mini-Lesson: What Is a Variable?

A **variable** is a place where a computer stores information.

Think of it like:

- a **jar** that can hold different numbers of marbles
- a **scoreboard** that keeps track of points
- a **labeled box** where you can put something in, take it out, or replace it
- a **sticky note** that you can erase and rewrite

A variable has **two important parts**:

1. **Its name** — what you call the container
2. **Its value** — what's inside the container right now

For example:

```
score = 0
temperature = 72
battery = 100
cookies = 12
```

These are all variables with different values. Notice how each variable has a name that tells you what it's storing. You shouldn't call it x or `thing` — that wouldn't help anyone! A good variable name should make sense, be easy to read, and match the idea it's tracking.

Variables Can Change

The whole point of a variable is that its value can change.

Just like:

- your age changes every year
- your piggy bank amount changes when you add coins
- your backpack weight changes depending on what's inside

Computers change variables with simple updates:

```
score = score + 1
temperature = temperature + 5
cookies = cookies - 1
battery = battery - 5
```

Every time something happens, the variable gets a new value.

Variables + Conditions

Variables become powerful when you use them in decisions. For example:

If score is 10, you win the game.

Or:

If battery is low, recharge.

The computer checks the variable's value and chooses what to do.

Variables + Loops

Variables also work beautifully with loops. For example:

Repeat 5 times:

```
Add 1 to score
```

Or:

Repeat until battery is 0:

```
Move forward
battery = battery - 1
```

Loops and variables together let computers keep track of progress.

Why Variables Matter

Variables let computers:

- remember information
- update information
- make decisions based on information
- track progress over time

Without variables, a computer would forget everything the moment it happened.

With variables, your programs become smarter, more flexible, and more interactive.

Your Turn

In the activities ahead, you'll:

- spot variables
- follow how they change
- rewrite instructions using variables
- fix broken variable logic
- and write your own variable-based rules

You're about to teach your computer how to **remember, update, and use information** — just like you do every day.

X

Activity Set A: Identify the Variable

Activity A1 — Counting Steps

Underline the **variable**.

Set `steps` to 0.

Helper Box		
Hint	*Common Mistake*	*What to Look For*
The variable is the name of the thing being stored.	Kids may underline the number instead of the variable name	steps

Activity A2 — Tracking Temperature

Underline the **variable**.

The `temperature` starts at 72.

Helper Box		
Hint	*Common Mistake*	*What to Look For*
Look for the word that holds the value.	Underlining the value (72).	temperature

☒

Activity A3 — Cookie Jar

Underline the **variable**.

Set `cookies` to 12.

Helper Box		
Hint	*Common Mistake*	*What to Look For*
Think: what is the computer keeping track of?	Kids may underline the action "set."	cookies

Activity A4 — Robot Battery

Underline the **variable**.

The robot's `battery` is 100.

Helper Box		
Hint	*Common Mistake*	*What to Look For*
The variable is the part that can change.	Underlining "robot."	battery

Activity A5 — Game Score

Underline the **variable**.

Set `score` to 0.

Helper Box		
Hint	*Common Mistake*	*What to Look For*
Scoreboards track changing numbers — so do variables.	Underlining the number 0.	score

Activity A6 — Water Level Sensor

Underline the **variable**.

The `waterLevel` starts at 50.

Helper Box		
Hint	*Common Mistake*	*What to Look For*
Look for the name of the thing being measured.	Kids may underline "starts."	waterLevel

Activity A7 — Backpack Weight

Underline the **variable**.

Set `weight` to 5.

Helper Box		
Hint	*Common Mistake*	*What to Look For*
Think about what changes when you add or remove items.	Underlining the number instead of the variable.	weight

Activity A8 — Drone Altitude

Underline the **variable**.

The drone's `altitude` is 20.

Helper Box		
Hint	*Common Mistake*	*What to Look For*
The variable is the thing the drone is keeping track of.	Underlining "drone."	altitude

Activity Set B: Track Changes to a Variable

Activity B1 — Piggy Bank

Follow the steps. Write the final value of `coins`.

1. Set `coins` to 3.
2. Add 2 to `coins`.
3. Add 1 to `coins`.

Helper Box		
Hint	*Common Mistake*	*What to Look For*
Update the value after each step.	Kids may only add the last change.	A final value of 6.

☒

Activity B2 — Jump Counter

Follow the steps. Write the final value of `jumps`.

1. Set `jumps` to 0.
2. Add 1 to `jumps`.
3. Add 1 to `jumps`.
4. Add 1 to `jumps`.

Helper Box		
Hint	*Common Mistake*	*What to Look For*
Count each `jump` one at a time.	Skipping a step or adding too many.	A final value of 3.

☐

Activity B3 — Water Bottle Tracker

Follow the steps. Write the final value of `water`.

1. Set `water` to 10.
2. Subtract 2 from `water`.
3. Subtract 1 from `water`.

Helper Box		
Hint	*Common Mistake*	*What to Look For*
Subtract means the value gets smaller.	Adding instead of subtracting.	A final value of 7.

☒

Activity B4 — Robot Battery

Follow the steps. Write the final value of `battery`.

1. Set `battery` to 100.
2. Subtract 10 from `battery`.
3. Subtract 10 from `battery`.
4. Add 5 to `battery`.

____

____

Helper Box		
Hint	*Common Mistake*	*What to Look For*
Keep track of each change in order.	Forgetting the final "add 5."	A final value of 85.

X

Activity B5 — Drone Altitude

Follow the steps. Write the final value of `altitude`.

1. Set `altitude` to 20.
2. Add 10 to `altitude`.
3. Subtract 5 from `altitude`.

Helper Box		
Hint	*Common Mistake*	*What to Look For*
Think of `altitude` like going up and down.	Mixing up the add and subtract steps.	A final value of 25.

☒

Activity B6 — Temperature Sensor

Follow the steps. Write the final value of `temperature`.

1. Set `temperature` to 70.
2. Add 2 to `temperature`.
3. Add 3 to `temperature`.
4. Subtract 1 from `temperature`.

Helper Box		
Hint	*Common Mistake*	*What to Look For*
Update the value after each step, not all at once.	Adding all numbers together without starting from 70.	A final value of 74.

☐ X

Activity B7 — Scoreboard

Follow the steps. Write the final value of `score`.

1. Set `score` to 0.
2. Add 5 to `score`.
3. Add 5 to `score`.
4. Subtract 3 from `score`.

Helper Box		
Hint	*Common Mistake*	*What to Look For*
Scoreboards go up and down depending on the game.	Forgetting the subtraction at the end.	A final value of 7.

☒

Activity B8 — Steps Tracker

Follow the steps. Write the final value of `steps`.

1. Set `steps` to 100.
2. Add 20 to `steps`.
3. Add 30 to `steps`.
4. Subtract 10 from `steps`.

Helper Box		
Hint	*Common Mistake*	*What to Look For*
Keep a running total as you go.	Adding all numbers without starting from 100.	A final value of 140.

Activity Set C: Using Variables in Real-World Scenarios

Activity C1 — Track the Temperature

Scenario: The classroom thermometer shows 68 degrees in the morning. By lunchtime, the temperature rises by 4 degrees. Later in the afternoon, it cools down by 2 degrees.

Variable name: `temperature`

Start at:

Increase by:

Decrease by:

Final value:

Helper Box		
Hint	*Common Mistake*	*What to Look For*
Take the numbers directly from the story.	Writing the *final* temperature instead of the *changes*.	A starting value of 68, then +4, then −2, and a final value of 70.

☐ X

Activity C2 — Counting Apples

Scenario: Riley has a basket of apples. They start with 5 apples. Their neighbor gives them 3 more. Later, they use 2 apples to make a snack.

Variable name: `apples`

Start at:

Add:

Subtract:

Final value:

Helper Box		
Hint	*Common Mistake*	*What to Look For*
This is a simple "add then subtract" pattern.	Kids sometimes subtract the wrong number or forget the order.	Start = 5, then +3, then −2, and finally 6.

X

Activity C3 — Steps on a Pedometer

Scenario: A pedometer starts the day at 0 steps. After recess, it shows 600 steps. After PE class, it increases by another 400 steps.

Variable name: `steps`

Start at:

Add after recess:

Final value:

Add after PE:

Helper Box		
Hint	*Common Mistake*	*What to Look For*
Both updates should increase the total.	Kids sometimes write the *new* total instead of the *amount added.*	Start = 0, then +600, then +400, and finally 1000.

☒

Activity C4 — Money in a Piggy Bank

Scenario: A piggy bank has $12 inside. You earn $4 for doing chores. Later, you spend $3 on a snack.

Variable name: `money`

Start with:

Earn:

Spend:

Final value:

Helper Box		
Hint	*Common Mistake*	*What to Look For*
Spending means subtracting.	Kids sometimes write "money = 0" after spending instead of subtracting.	Start = 12, then +4, then −3, finally 13.

Activity C5 — Snack Tracker

Scenario: Jordan starts with 10 crackers. They eat 3 during snack time and 2 more after school. Track how many crackers remain.

Rewrite using a variable: Pick a descriptive variable name, a starting value, updates, and a final value.

Helper Box		
Hint	*Common Mistake*	*What to Look For*
A good variable name might be `crackers`.	Kids sometimes write the final total instead of the updates.	A clear starting assignment, two update statements, and a final value.

☒

Activity C6 — Water Bottle

Scenario: A water bottle starts with 20 ounces of water. You drink 5 ounces, then refill it with 8 ounces.

Rewrite using a variable: Pick a descriptive variable name, a starting value, updates, and a final value.

Helper Box		
Hint	*Common Mistake*	*What to Look For*
Think: start → subtract → add.	Kids sometimes reverse the order of the updates.	A starting value, two updates in the correct order, and a final value.

Activity C7 — Art Supply Tracker

Scenario: You are tracking sheets of colored paper. You begin with 6 sheets, receive 4 more, then use 2 sheets to make a poster.

Rewrite using a variable: Pick a descriptive variable name, a starting value, updates, and a final value.

Helper Box		
Hint	*Common Mistake*	*What to Look For*
Pick a variable name that matches the story.	Using a name like `cat` instead of something that matches the scenario.	A descriptive variable name, correct updates, and a correct final value.

☒

Activity C8 — Rainfall Tracker

Scenario: A rain gauge starts with 0 inches. It collects 2 inches in the morning and 4 inches in the afternoon.

Rewrite using a variable: Pick a descriptive variable name, a starting value, updates, and a final value.

Helper Box		
Hint	*Common Mistake*	*What to Look For*
This one only increases — no subtraction.	Kids sometimes forget to start at 0.	A starting value, two correct additions, and a final value.

Activity Set D: Debugging Variables

Activity D1 — The Cookie Counter

Scenario: Sam is counting cookies. They start with 10 cookies. They get 3 more from their friend, then they eat 2.

Broken Instructions (There is 1 mistake):

```
cookies = 10
cookies = cookies + 3
cookies = 5
cookies = cookies - 2
```

Rewrite the corrected instructions and the final value of the variable:

____

____

____

____

`cookies =` ___________________________________

Helper Box		
Hint	*Common Mistake*	*What to Look For*
Look for a line that resets the variable instead of updating it.	Kids often think cookies = 5 means "add 5." Instead, it sets the total cookies to 5.	A clean sequence: start → add → subtract.

Activity D2 — The Backpack Weight Tracker

Scenario: A backpack weighs 4 pounds. You add a 2-pound book, then remove a 3-pound binder.

Broken Instructions (There are 2 mistakes):

```
weight = 5
weight = weight + 2
weight = weight - 3
weight = weight - 4
```

Rewrite the corrected instructions and the final value of the variable:

____

____

____

____

`weight =` ____________________

Helper Box		
Hint	*Common Mistake*	*What to Look For*
Check the starting value and each update against the story.	Fixing only one mistake: the incorrect starting value or the extra subtraction.	Start = 4, then +2, then −3, finally 3.

Activity D3 — The Scoreboard Mix-Up

Scenario: A game scoreboard starts at 0. You score 5 points, then 2 more.

Broken Instructions (There is 1 mistake):

```
score = 0
score = score + 5
score = score - 2
```

Rewrite the corrected instructions and the final value of the variable:

____

____

____

____

`score =` ______________________________

Helper Box		
Hint	*Common Mistake*	*What to Look For*
Compare each update with what happens in the scenario.	Changing the final value while leaving `score = score - 2` unchanged.	The final value should be 7.

☒

Activity D4 — The Water Bottle Tracker

Scenario: A water bottle has 20 ounces. You drink 5 ounces, then refill it with 8 ounces.

Broken Instructions (There is 1 mistake):

```
water = 20
water = water - 5
water = water + 6
```

Rewrite the corrected instructions and the final value of the variable:

____

____

____

____

`water =` ______________________________

Helper Box		
Hint	*Common Mistake*	*What to Look For*
Only include the updates that match the story.	Calculating the intended final value while leaving + 6 unchanged.	Start = 20, then −5, then +8, finally 23.

☒

Activity D5 — The Toy Car Collector

Scenario: You have 6 toy cars. You find 4 more, then give 2 to a friend.

Broken Instructions (There is 1 mistake):

```
cars = 6
cars = cars - 2
```

Rewrite the corrected instructions and the final value of the variable:

___

___

___

___

`cars =` ___

Helper Box		
Hint	*Common Mistake*	*What to Look For*
Compare each line to the story.	Subtracting the two given-away cars without adding the four newly found cars.	Start → add → subtract.

☒

Activity D6 — The Step Counter Glitch

Scenario: Your step counter starts at 0. You walk 600 steps at recess and 200 more after school.

Broken Instructions (There is 1 mistake):

```
steps = 0
steps = steps + 600
steps = steps + 400
```

Rewrite the corrected instructions and the final value of the variable:

____

____

____

____

`steps =` ____________________________________

Helper Box		
Hint	*Common Mistake*	*What to Look For*
Compare each number with the scenario.	Using the 400-step update from the earlier pedometer activity.	Start → add → add.

X

Activity D7 — The Piggy Bank Problem

Scenario: Your piggy bank has 12 dollars. You earn 4 dollars, then spend 3 dollars.

Broken Instructions (There is 1 mistake):

```
money = 12
money = money + 4
mony = money - 3
```

Rewrite the corrected instructions and the final value of the variable:

____

____

____

____

`money =` __________________________

Helper Box		
Hint	*Common Mistake*	*What to Look For*
Variable names must match exactly.	Calculating 13 while leaving the misspelled variable name unchanged.	Change `mony` to `money`. The final value of `money` is 13.

☒

Activity D8 — The Rain Gauge Reset

Scenario: A rain gauge starts at 0 inches. It collects 2 inches in the morning and 4 inches in the afternoon.

Broken Instructions (There is 1 mistake):

```
rain = 0
rain = rain + 2
rain = rain + 4
rain = 0
```

Rewrite the corrected instructions and the final value of the variable:

____

____

____

____

`rain =` ____

Helper Box		
Hint	*Common Mistake*	*What to Look For*
The variable should change only when something happens in the scenario.	Removing one of the additions instead of the final reset.	Remove the final `rain = 0`. The final value is 6.

Activity Set E: Variable Challenges

Activity E1 — Two Variables at the Animal Shelter

Scenario: You are helping count animals at an animal shelter. There are 6 puppies and 4 kittens. Two puppies get adopted, and 3 new kittens arrive.

Your task: Create two variables (one for puppies and one for kittens) and write the starting values and updates.

Rewrite using variables:

Helper Box		
Hint	*Common Mistake*	*What to Look For*
Use two separate variables so the numbers don't get mixed.	Kids sometimes add kitten numbers to the puppy variable or vice-versa.	Two clear variable names and correct updates for each.

☒

Activity E2 — The Lemonade Stand Tracker

Scenario: You start the day with 20 cups of lemonade. You sell 7 cups in the morning, refill with 10 cups at lunch, then sell 8 more in the afternoon.

Your task: Choose a variable name and write the starting value and all updates.

Rewrite using a variable:

____

____

____

____

____

____

Helper Box		
Hint	*Common Mistake*	*What to Look For*
Think: start → subtract → add → subtract.	Kids sometimes forget the refill step.	A variable name that matches the story and all three updates.

X

Activity E3 — The Arcade Ticket Machine

Scenario: You start with 0 tickets. You win 15 tickets on your first game, lose 5 tickets on a trade, then win 12 more.

Your task: Write the variable updates in order.

Rewrite using a variable:

Helper Box		
Hint	*Common Mistake*	*What to Look For*
Losing tickets means subtracting.	Kids sometimes write the final total instead of the updates.	Start at 0 and include all three changes.

☒

Activity E4 — Library Book Tracker

Scenario: You are tracking how many library books you have checked out. You start with 3 books, check out 2 more, return 1, then check out 4 more.

Your task: Pick a variable name and write the starting value and updates.

Rewrite using your variable:

____

____

____

____

____

____

Helper Box		
Hint	*Common Mistake*	*What to Look For*
A good variable name might be books or `libraryBooks`.	Kids sometimes forget that returning means subtraction.	A meaningful variable name and all updates in order.

X

Activity E5 — Recycling Center Inventory

Scenario: At a recycling center, 10 plastic bottles and 8 aluminum cans are waiting to be sorted. Workers sort and remove 4 bottles and 3 cans from the unsorted inventory. Later, a truck adds 6 bottles and 2 cans to the unsorted inventory.

Your task: Create two variables and show all updates.

Rewrite using variables:

____

____

____

____

____

____

Helper Box		
Hint	*Common Mistake*	*What to Look For*
Keep the bottle updates and can updates separate.	Kids sometimes combine the numbers into one variable.	Two variables with correct starting values and updates.

X

Activity E6 — The Daily Step Challenge

Scenario: You start the day with 0 steps. You walk 1,200 steps before lunch, 900 steps after lunch, and 600 steps in the evening.

Your task: Write the starting value and all updates using a variable.

Rewrite using a variable:

Helper Box		
Hint	*Common Mistake*	*What to Look For*
All updates are additions.	Kids sometimes write the total instead of the updates.	Start at 0 and include all three additions.

X

Activity E7 — Savings Tracker

Scenario: You are saving money in a jar. You start with 5 dollars, earn 8 dollars doing chores, spend 3 dollars on a snack, then earn 6 more dollars.

Your task: Pick a variable name and write the starting value and updates.

Rewrite using your variable:

Helper Box		
Hint	*Common Mistake*	*What to Look For*
Think about which steps add and which subtract.	Kids sometimes reverse the order of the updates.	A clear variable name and all four updates.

Activity E8 — Challenge: Two Variables, Two Updates

Scenario: You are tracking two things: pages read and minutes spent reading. You start with 0 pages and 0 minutes. You read 12 pages in 15 minutes, then 8 more pages in 10 minutes.

Your task: Create two variables and write all updates.

Rewrite using variables:

Helper Box		
Hint	*Common Mistake*	*What to Look For*
Use one variable for pages and one for minutes.	Kids sometimes mix the page numbers and minute numbers.	Two variables with correct starting values and two updates each.

CERTIFICATE OF ACHIEVEMENT

Variable Virtuoso

This certificate is awarded to ______________________________ for completing all activities in **Chapter 6: Variables** of *Think Like a Coder*.

___________________________ Helper Signature

___________________________ Date

LOGIC LAB BOOKS

Chapter 7: Functions

Patterns Become Powerful

Introduction: Functions

Have you ever wished you could press a button and have something happen automatically? Maybe a button that cleans your room, or one that makes your favorite snack appear? Computers have something just like that — they're called *functions*.

A function is like a tiny machine inside your program. You give it a name, you tell it what job it should do, and then you can use it again and again without rewriting all the steps. It's like having your own helper robot that follows your instructions perfectly every time.

In this chapter, you'll learn how to create functions, how to use them, and why programmers love them so much. By the end, you'll be able to build your own little machines that make your code cleaner, faster, and easier to understand.

Mini-Lesson: What Is a Function?

A function is a named set of instructions. Instead of writing the same steps over and over, you can put those steps into a function and give it a name. Then, whenever you need those steps, you just "call" the function.

It's like writing a recipe once and then using it every time you want to make the same dish.

Real-Life Example

Imagine you have a morning routine:

1. Brush teeth
2. Wash face
3. Eat breakfast

You do these steps every day. Instead of listing them out each time, you could name the whole routine:

```
morningRoutine
```

Now, whenever someone asks what you do in the morning, you can just say "I do my `morningRoutine`," and they know it includes all three steps.

That's exactly how functions work in programming.

A Simple Coding-Style Example

Let's use pseudocode to show an example. Pseudocode is pretend code. It shows the steps a computer would follow, and it's written in easy-to-read language instead of any particular programming language. It helps you understand the idea of the code before you learn the exact symbols and rules. Here's what a function might look like in pseudocode:

```
function sayHello
    print "Hello!"
end function
```

And here's how you *use* it:

```
sayHello
sayHello
sayHello
```

When you *use* a function, programmers say you *call* it. Each time you call a function, the computer runs the instructions inside the function.

Output:

```
Hello!
Hello!
Hello!
```

Why Functions Matter

Functions help programmers:

- Avoid repeating the same code
- Keep programs organized
- Make changes in one place instead of many
- Build bigger, more powerful programs

If you ever find yourself writing the same steps more than once, that's a clue: **It's time to make a function!**

Functions Can Use Variables

Functions can work with variables you already know how to use. For example:

```
function cheer
    print "You can do it!"
    print "Your score is " + score
end function
```

When you call the function, it uses the current value of the variable.

A Function That Does More Than One Thing

Functions can contain as many steps as you need:

```
function makeSandwich
    get bread
    add peanutButter
    add jelly
    put together
end function
```

Calling `makeSandwich` runs all those steps in order.

Calling a Function

When you "call" a function, you're telling the computer:

"Run the steps inside this function right now."

It's like pressing the button on your tiny machine.

Functions Don't Run Until You Call Them

When you write a function, you're *building* a tiny machine — and building the machine doesn't make it run. The steps inside the function won't happen until you *call* the function by using its name. This is one of the biggest surprises for new coders! They write all the steps, press Run, and wonder why nothing happens. The reason is simple: the computer is waiting for the call. Once you call the function, the computer knows it's time to run all the steps inside your tiny machine.

How to Read a Function

When you see a function, ask three questions:

1. What is the function's name?
2. Which instructions are inside it?
3. Where is the function called?

Follow the program from top to bottom. Each time you reach a function call, run every instruction inside the function in order. Then continue with the next line of the program.

Putting It All Together

Functions are one of the most powerful tools in programming. They let you bundle steps together, give those steps a name, and use them whenever you need them. Instead of rewriting the same instructions again and again, you can just call the function and let your tiny machine do the work for you.

As you learn more about functions, you'll start to see them everywhere — in games, apps, websites, and even in the code you write yourself. And now you're ready to start using functions just like a real programmer: creating them, calling them, and building bigger ideas out of smaller pieces.

Activity Set A: Understanding Functions

Activity A1 — What Does This Function Do?

```
function greet
    print "Hello!"
    print "Welcome!"
end function
```

When you call the function `greet`, what will the computer print?

Helper Box		
Hint	*Common Mistake*	*What to Look For*
Look at the steps inside the function.	Kids sometimes think the function name prints by itself.	Both print lines in order.

Activity A2 — Identify the Function Name

```
function makeBed
    straightenBlanket
    fluffPillow
end function
```

What is the name of this function?

Helper Box		
Hint	*Common Mistake*	*What to Look For*
The name comes right after the word function.	Kids sometimes write one of the steps instead of the name.	A single word that describes the job.

Activity A3 — What Happens When You Call It?

```
function cheer
    print "You can do it!"
end function

cheer
cheer
```

How many times will the message print?

____

____

Helper Box		
Hint	*Common Mistake*	*What to Look For*
Each call runs the whole function.	Kids count the print line instead of the calls.	Two calls = two messages.

Activity A4 — Match the Function to Its Job

```
function jump
    print "Jump!"
end function

function wave
    print "Wave!"
end function

function cheer
     print "Cheer!"
end function
```

Match each call to what it makes the computer print.

`jump`	Cheer!
`wave`	Jump!
`cheer`	Wave!

Helper Box		
Hint	*Common Mistake*	*What to Look For*
The function name tells you what it does.	Kids mix up the names because the functions look similar.	One prints "Jump!", one prints "Wave!", and one prints "Cheer!".

Activity A5 — Which One Is the Function Call?

```
function startGame
    print "Ready!"
end function

startGame
```

Circle the line that *calls* the function.

Helper Box		
Hint	*Common Mistake*	*What to Look For*
The call is the line *outside* the function.	Kids circle the function name inside the definition.	The line with just the function name.

Activity A6 — What Does This Function Print?

```
function showScore
    print "Your score is:"
    print score
end function

score = 12
showScore
```

What two lines will the computer print?

Helper Box		
Hint	*Common Mistake*	*What to Look For*
The function uses the current value of the variable.	Kids forget to include the variable's value.	The text line, then the number.

Activity A7 — Does This Call the Function?

```
function dance
    print "Spin!"
end function
```

Circle the line that calls the function.

```
spin
```

```
dance
```

Helper Box		
Hint	*Common Mistake*	*What to Look For*
The call uses the function's name.	Kids sometimes choose the action instead of the function name.	The line that matches the function name exactly.

Activity A8 — What's Inside the Function?

Function:

```
function cleanRoom
    pickUpToys
    makeBed
    vacuum
end function
playGame
```

List the three steps this function does.

Helper Box		
Hint	*Common Mistake*	*What to Look For*
Look between the `function` and `end function` lines.	Kids include the function name or `playGame` as a step.	Only the actions inside the function.

Activity Set B: Calling Functions

Example

Here is a function.

```
function greet
     print "Hello!"
end function
greet
```

What will the computer print when you call `greet`?

Hello!

Activity B1 — Calling a Simple Function

```
function sayHi
    print "Hi there!"
end function
sayHi
```

What will the computer print when you call `sayHi`?

Helper Box		
Hint	*Common Mistake*	*What to Look For*
Look at the steps inside the function.	Kids sometimes think the function name prints by itself.	The exact text inside the print line.

Activity B2 — Calling a Function More Than Once

```
function cheer
    print "Go team!"
end function


cheer
cheer
cheer
```

How many times will the message print?

____

____

Helper Box		
Hint	*Common Mistake*	*What to Look For*
Each call runs the whole function.	Kids count the print line instead of the calls.	Three calls = three messages.

Activity B3 — Calling a Function With Multiple Steps

```
function warmUp
    stretchArms
    stretchLegs
    jogInPlace
end function

warmUp
```

List the three steps the computer will run when you call `warmUp`.

Helper Box		
Hint	*Common Mistake*	*What to Look For*
Look between the `function` and `end function` lines.	Kids include the function name as a step.	Only the actions inside the function.

Activity B4 — Calling a Function After Setting a Variable

```
function showName
    print "Your name is:"
    print name
end function

name = "Taylor"
showName
```

What two lines will the computer print?

____

____

____

Helper Box		
Hint	*Common Mistake*	*What to Look For*
The function uses the current value of the variable.	Kids forget to include the variable's value.	The text line, then the name.

Activity B5 — Which Line Calls the Function?

```
function startGame
    print "Loading..."
end function

print "Welcome!"
startGame
print "Good luck!"
```

Which line calls the function?

Helper Box		
Hint	*Common Mistake*	*What to Look For*
The call uses only the function's name.	Kids choose the print lines instead.	The line that matches the function name exactly.

Activity B6 — Calling Two Different Functions

```
function jump
    print "Jump!"
end function

function clap
    print "Clap!"
end function

jump
clap
jump
```

Write the three messages the computer will print in order.

Helper Box		
Hint	*Common Mistake*	*What to Look For*
Follow the calls in order.	Kids mix up the order of the messages.	Jump → Clap → Jump.

Activity B7 — Calling a Function Inside a Sequence

```
function ringBell
    print "Ding!"
end function

print "Class is starting"
ringBell
print "Take your seats"
```

What will the computer print, in order?

Helper Box		
Hint	*Common Mistake*	*What to Look For*
The function call happens in the middle.	Kids put the bell sound at the beginning or the end.	The function runs exactly where it appears.

Activity B8 — What Happens When You Call It Twice?

```
function showScore
    print "Score:"
    print score
end function

score = 10
showScore
score = 15
showScore
```

What will the computer print each time you call `showScore`?

Helper Box		
Hint	*Common Mistake*	*What to Look For*
The function uses the *current* value of the variable each time.	Kids think the function always uses the first value.	Score changes between calls.

Activity Set C: Writing Your Own Functions

Activity C1 — Write a Function That Says Hello

Write a function named `sayHello` that prints two friendly messages.

```
function sayHello
________________________________________

________________________________________

________________________________________
end function
________________________________________
```

Helper Box		
Hint	*Common Mistake*	*What to Look For*
Use two print lines.	Kids forget to include both steps.	A function name and two clear print statements.

Activity C2 — Write a Function for a Morning Routine

Create a function named `morningRoutine` with three steps you do every morning.

```
function morningRoutine
_______________________________________________
_______________________________________________
_______________________________________________
end function
_______________________________________________
```

Helper Box		
Hint	*Common Mistake*	*What to Look For*
Think of real steps: brush teeth, eat breakfast, etc.	Kids write the function name as a step.	Three real actions inside the function.

Activity C3 — Write a Function and Call It

Write a function named `cheer` that prints a short cheer. Then write a line that *calls* the function.

```
function cheer

end function
```

Helper Box		
Hint	*Common Mistake*	*What to Look For*
The call is just the function name on its own line.	Kids put the call inside the function.	Function first, call second.

Activity C4 — Write a Function With a Variable

Write a function named `showLevel` that prints the current level. Use the variable `level` inside the function.

```
level = 3
function showLevel

end function
```

Helper Box		
Hint	*Common Mistake*	*What to Look For*
One line should print text, the next should print the variable.	Kids write the number instead of the variable name.	A print line with text and a print line with the variable.

Activity C5 — Create a Function for a Pet

Choose a pet (real or imaginary!) and write a function that describes how to take care of it.

Give the function a descriptive name that explains how it cares for your pet.

```
function ____________________

____________________
____________________
____________________
end function
____________________
```

Helper Box		
Hint	*Common Mistake*	*What to Look For*
Use a name that describes the function's purpose.	Kids forget to fill in the function name.	Three steps that make sense for the chosen pet.

Activity C6 — Write a Function With Two Actions

Write a function named `startRace` that prints "Ready..." and then prints "Go!"

```
function startRace
______________________________
______________________________
______________________________
end function
______________________________
```

Helper Box		
Hint	*Common Mistake*	*What to Look For*
Use two print lines in order.	Kids combine both messages into one line.	Two separate print statements.

Activity C7 — Write a Function and Use It Twice

Write a function named `clapHands` that prints "Clap! Clap!" Then call the function two times.

```
function clapHands

end function
```

Helper Box		
Hint	*Common Mistake*	*What to Look For*
The calls go *after* the function.	Kids put both calls inside the function.	One function, two calls.

Activity C8 — Create Your Own Function

Make up your own function! Choose a name and write 2–4 steps inside.

```
function ____________________________
_____________________________________
_____________________________________
_____________________________________
_____________________________________
end function
```

Helper Box		
Hint	*Common Mistake*	*What to Look For*
Pick a name that matches the job.	Kids write steps that don't relate to the function name.	A clear name and matching steps.

Activity Set D: Debugging Functions

Activity D1 — The Function Is Never Called

```
function sayHello
    print "Hello!"
end function
```

Why does nothing print when you run the program?

Helper Box		
Hint	*Common Mistake*	*What to Look For*
Building a function doesn't make it run.	Kids think writing the function is enough.	The function needs a call.

Activity D2 — The Function Name Doesn't Match

```
function startGame
    print "Ready!"
end function

beginGame
```

Why does the function not run?

___

___

___

Helper Box		
Hint	*Common Mistake*	*What to Look For*
The call must match the function name exactly.	Kids mix up similar names.	`startGame` vs. `beginGame`.

Activity D3 — A Step Is Outside the Function

```
function warmUp
    stretchArms
end function
stretchLegs
```

What is wrong with this function?

Helper Box		
Hint	*Common Mistake*	*What to Look For*
All steps should be inside the function.	Kids forget to include all steps before the end.	`stretchLegs` is outside the function.

Activity D4 — Missing the End of the Function

```
function cheer
    print "Go team!"
print "You can do it!"
```

What is missing from this function?

Helper Box		
Hint	*Common Mistake*	*What to Look For*
Every function needs a clear ending.	Kids forget to write end `function`.	The function never closes.

Activity D5 — Wrong Variable Name

```
function showScore
    print "Score:"
    print points
end function

score = 12
showScore
```

Why does this function not show the correct number?

Helper Box		
Hint	*Common Mistake*	*What to Look For*
The variable inside the function must match the one you set.	Kids mix up score and points.	The function prints the wrong variable.

Activity D6 — The Call Is Inside the Function

```
function ringBell
    print "Ding!"
    ringBell
end function
```

Why is this a problem?

Helper Box		
Hint	*Common Mistake*	*What to Look For*
Move ringBell outside the function so it only runs once.	Kids think the call belongs inside the function.	The call should be *after* the function.

Activity D7 — Steps in the Wrong Order

```
function makeSandwich
    putTogether
    addJelly
    addPeanutButter
    getBread
end function
```

What is wrong with the order of the steps?

Helper Box		
Hint	*Common Mistake*	*What to Look For*
Think about how you actually make a sandwich.	Kids write steps backwards.	The steps should go from start to finish.

Activity D8 — Extra Step That Doesn't Belong

```
function startRace
    print "Ready..."
    print "Go!"
    print "Game Over"
end function
```

Which line does not belong in this function?

Helper Box		
Hint	*Common Mistake*	*What to Look For*
Think about what the function is supposed to do.	Kids include steps that don't match the function name.	"Game Over" doesn't fit the job.

Activity Set E: Function Puzzles

Activity E1 — Two Calls, Two Different Values

```
function showLevel
    print "Level:"
    print level
end function

level = 2
showLevel
level = 5
showLevel
```

What will the computer print each time the function is called?

First call:

Second call:

Helper Box		
Hint	*Common Mistake*	*What to Look For*
The function uses the *current* value of the variable.	Kids think the function always uses the first value.	Two different outputs.

Activity E2 — A Function Inside a Sequence

```
function cheer
    print "You can do it!"
end function

print "Welcome"
cheer
print "Let's begin"
```

Write the three lines the computer will print, in order.

Helper Box		
Hint	*Common Mistake*	*What to Look For*
The function runs exactly where it appears.	Kids put the cheer at the end.	Welcome → You can do it! → Let's begin.

Activity E3 — Fix the Function Name

```
function startTimer
    print "Timer started!"
end function

startTime
```

Rewrite the call so the function runs correctly.

Helper Box		
Hint	*Common Mistake*	*What to Look For*
The call must match the function name exactly.	Kids copy the wrong name.	`startTimer`

Activity E4 — Write a Function With Three Steps

Write a function named `waterPlants` with three steps you do to care for plants.

```
function waterPlants
_______________________________
_______________________________
_______________________________
end function
_______________________________
```

Helper Box		
Hint	*Common Mistake*	*What to Look For*
Think: watering can, soil, sunlight, etc.	Kids write the function name as a step.	Three real actions.

Activity E5 — Predict the Output

```
function showScore
    print "Score:"
    print score
end function

score = 10
showScore
score = score + 4
showScore
```

What will the computer print each time?

First call:

Second call:

Helper Box		
Hint	*Common Mistake*	*What to Look For*
The variable changes between calls.	Kids forget to add 4.	Score 10, then score 14.

Activity E6 — Write a Function and Use It Twice

Write a function named `ringBell` that prints "Ding!" Then call it two times.

```
function ringBell
________________________________________

end function
________________________________________
________________________________________
________________________________________
```

Helper Box		
Hint	*Common Mistake*	*What to Look For*
The calls go after the function.	Kids put both calls inside the function.	One function, two calls.

Activity E7 — Debug the Function

This function should pack school materials and then run once. It has two mistakes.

Rewrite it so every step is inside the function and the call uses the correct name.

```
function packBackpack
     addHomework
end function
addNotebook
zipBackpack
packBag
```

Helper Box		
Hint	*Common Mistake*	*What to Look For*
Compare the function boundaries and the function-call name.	Kids may fix only one of the two mistakes.	All three packing steps inside packBackpack, followed by a matching function call.

Activity E8 — Create Your Own Challenge Function

Make up your own function! Choose a name and write 3–4 steps inside. Then write one line that calls your function.

```
function ____________________
____________________
____________________
____________________
____________________
end function
____________________
____________________
```

Helper Box		
Hint	*Common Mistake*	*What to Look For*
Pick a name that matches the job.	Kids forget to write the call.	A clear name, matching steps, and one call.

CERTIFICATE OF ACHIEVEMENT

Function Force

This certificate is awarded to ______________________________ for completing all activities in **Chapter 7: Functions** of *Think Like a Coder*.

___________________________ Helper Signature

___________________________ Date

LOGIC LAB BOOKS

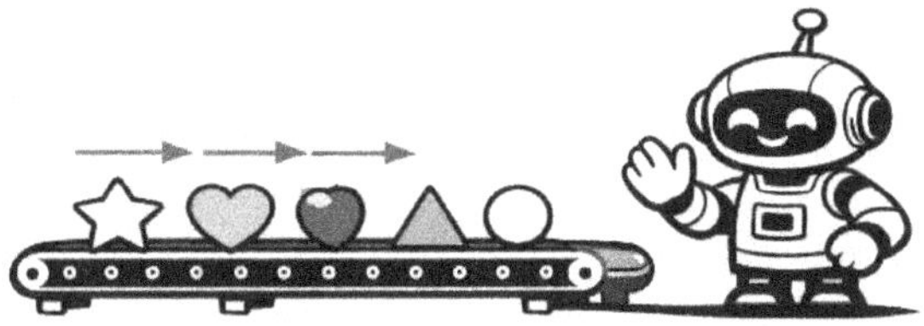

Chapter 8: Lists

Order Creates Meaning

Introduction: Lists

Have you ever tried to carry too many things at once? Maybe you had your backpack, a lunchbox, a water bottle, and a jacket all in your hands. It's a lot to keep track of! Wouldn't it be easier if you could put everything into one bag?

That's exactly what a **list** does in programming. A list lets you store a bunch of related items in one place so you can use them, change them, or look at them whenever you need. Instead of keeping track of ten separate things, you can keep track of *one* list.

In this chapter, you'll learn how to create lists, read items from lists, and use lists to make your programs smarter and more organized.

Mini-Lesson: What Is a List?

A List Is a Group of Items

A list is a collection of things stored together under one name. You can think of it like:

- A playlist
- A list of chores
- A list of your favorite games

Instead of storing each item separately, you store them all together.

Example in pseudocode:

```
games = ["Minecraft", "Roblox", "Zelda"]
```

Now the variable `games` holds *three* items.

Why Lists Are Useful

Lists help programmers:

- Keep related items together
- Add or remove items easily
- Use loops to go through each item
- Build bigger programs with less code

If you ever find yourself writing lots of separate variables like:

```
code
game1
game2
game3
```

That's a clue: "It's time to use a list!"

Reading Items From a List

Each item in a list has a position called an **index**. In most programming languages, the first item is at index **0**.

Example:

```
pets = ["dog", "cat", "hamster"]
pets[0] is "dog"
pets[1] is "cat"
pets[2] is "hamster"
```

This lets you pick out exactly the item you want.

Changing Items in a List

You can update items in a list just like you update a variable.

```
pets[1] = "parrot"
```

Now the list is:

```
["dog", "parrot", "hamster"]
```

Adding Items to a List

Lists can grow!

```
pets.append("turtle")
```

Now the list has four items: dog, parrot, hamster, turtle.

Using Lists With Loops

Lists and loops are best friends.

A loop can go through each item in a list:

```
for each pet in pets
    print pet
end loop
```

This prints every pet in the list, one at a time. When you loop through a list, you get to choose a name for the variable that is used to access each item from the list. The name should make sense — it should tell you what the item *is*. If your list is `pets`, naming the item `pet` helps you understand that the variable is holding a kind of pet. If you named the variable p or `something`, it would be harder to remember that the variable is holding a kind of pet. Clear names make your code easier to read, easier to debug, and easier to think about.

Lists Don't Do Anything Until You Use Them

Just like functions don't run until you call them, lists don't do anything until you *use* them. Creating a list only stores the items — it doesn't print them or change them automatically.

A list is a container. You decide when to open it and what to do with the items inside.

Putting It All Together

Lists help you organize information, store groups of items, and use loops more powerfully. Once you learn how to work with lists, you'll be able to build programs that sort things, track things, and handle lots of data at once — just like real programmers do.

And now you're ready to start using lists to make your code smarter, cleaner, and more fun.

Activity Set A: Understanding Lists

Activity A1 — What Is in This List?

```
pets = ["dog", "cat", "hamster"]
```

Write all three items in the list.

Helper Box		
Hint	*Common Mistake*	*What to Look For*
Look between the square brackets.	Kids include the variable name as an item.	Only the three animals.

Activity A2 — Match Each Item to Its Index

```
colors = ["red", "blue", "green"]
```

Write each item next to its index.

Index 0 =

Index 1 =

Index 2 =

Helper Box		
Hint	*Common Mistake*	*What to Look For*
Index 0 is the first item, 1 is the second, and 2 is the third.	Kids think the first item is index 1.	0 . → red, 1 → blue, 2 → green.

Activity A3 — Identify the Second Item

```
snacks = ["chips", "apple", "granola bar"]
```

What is the item at index 1?

Helper Box		
Hint	*Common Mistake*	*What to Look For*
Index 1 is the second item.	Kids count starting at 1 instead of 0.	The item in the middle.

Activity A4 — How Many Items Are in the List?

```
games = ["Minecraft", "Roblox", "Zelda", "Mario Kart"]
```

How many items are in this list?

Helper Box		
Hint	*Common Mistake*	*What to Look For*
Count the items, not the indexes.	Kids count the highest index instead of the number of items.	Four items.

Activity A5 — Match the Index to the Item

```
fruits = ["banana", "orange", "grape"]
```

Draw a line connecting each Index to its value in the list.

Index 0	grape
Index 1	orange
Index 2	banana

Helper Box		
Hint	*Common Mistake*	*What to Look For*
Index numbers go in order from left to right.	Kids reverse the order.	banana → orange → grape.

Activity A6 — Which List Is Correct?

Option A

```
pets = "dog", "cat", "fish"
```

Option B

```
pets = ["dog", "cat", "fish"]
```

Which option shows a list written correctly? How do you know?

Helper Box		
Hint	*Common Mistake*	*What to Look For*
Lists use square brackets.	Kids choose the version with commas but no brackets.	The list with brackets.

Activity A7 — What Is the Index of the Last Item?

```
books = ["Magic Tree House", "Dog Man", "Wings of Fire"]
```

What is the index of the last item ("Wings of Fire")?

Helper Box		
Hint	*Common Mistake*	*What to Look For*
Remember the indexes start at 0.	Kids think index 2 means the second item.	The last book in the list.

Activity A8 — Create Your Own List

Create a list of three favorite foods. Write the list using square brackets.

Helper Box		
Hint	*Common Mistake*	*What to Look For*
Separate items with commas.	Kids forget the brackets or write only one item.	Three items inside square brackets.

Activity Set B: Reading and Changing Items

Activity B1 — Read an Item by Index

```
animals = ["dog", "cat", "turtle", "bird"]
```

What is the item at index 2?

Helper Box		
Hint	*Common Mistake*	*What to Look For*
Index 2 is the *third* item.	Kids count starting at 1.	"turtle".

Activity B2 — Change an Item in the List

```
colors = ["red", "blue", "green"]
colors[1] = "purple"
```

Write the new list after the change.

Helper Box		
Hint	*Common Mistake*	*What to Look For*
Only the item at index 1 changes.	Kids rewrite the whole list incorrectly.	`["red", "purple", "green"]`

Activity B3 — Add an Item to the End

```
snacks = ["chips", "apple"]
snacks.append("cookie")
```

Write the updated list.

Helper Box		
Hint	*Common Mistake*	*What to Look For*
append adds to the *end*.	Kids replace an item instead of adding one.	Three items, with "cookie" last.

Activity B4 — Replace the First Item

```
pets = ["hamster", "cat", "dog"]
pets[0] = "parrot"
```

What is the new first item in the list?

____

Helper Box		
Hint	*Common Mistake*	*What to Look For*
Index 0 is the first position.	Kids change the wrong item.	"parrot".

Activity B5 — Read Two Items

```
games = ["Minecraft", "Roblox", "Zelda", "Mario Kart"]
```

What are the items at index 1 and index 3?

Index 1: ______________________________

Index 3: ______________________________

Helper Box		
Hint	*Common Mistake*	*What to Look For*
Count from left to right starting at 0.	Kids shift everything by one.	"Roblox" and "Mario Kart".

Activity B6 — Change One Item, Leave the Rest

```
books = ["Magic Tree House", "Dog Man", "Wings of Fire"]
books[2] = "Captain Underpants"
```

Write the new list.

Helper Box		
Hint	*Common Mistake*	*What to Look For*
Only index 2 changes.	Kids rewrite all three items incorrectly.	Last item becomes “Captain Underpants”.

Activity B7 — Add an Item, Then Read One

```
foods = ["pizza", "tacos"]
foods.append("salad")
```

After adding the new item, what is the item at index 2?

Helper Box		
Hint	*Common Mistake*	*What to Look For*
Append adds to the end.	Kids think the new item goes in the middle.	“salad”.

Activity B8 — Create and Change Your Own List

Create a list of three favorite animals, then change the item at index 1 to something new.

```
animals =
```

Change the item at index 1:

```
animals[1] =
```

Write the new complete list:

```
animals =
```

Helper Box		
Hint	*Common Mistake*	*What to Look For*
The list should have exactly three items.	Kids forget to rewrite the list before changing it.	A valid list and a correct update.

Activity Set C: Using Lists With Loops

Example

Here is a list with a loop.

```
animals = ["dog", "cat", "turtle"]

for each animal in animals
    print animal
end loop
```

What will the computer print when you run the loop?

dog

cat

turtle

Activity C1 — Print Every Item in the List

```
pets = ["dog", "cat", "hamster"]

for each pet in pets
    print pet
end loop
```

What three words will the computer print?

Helper Box		
Hint	*Common Mistake*	*What to Look For*
The loop prints each item in order.	Kids print the list name instead of the items.	dog → cat → hamster.

Activity C2 — Loop Through a List of Numbers

```
scores = [3, 5, 7]

for each score in scores
    print score
end loop
```

Write the three numbers the loop will print.

____

____

____

Helper Box		
Hint	*Common Mistake*	*What to Look For*
Loops go through the list from start to finish.	Kids skip the middle item.	3, 5, 7.

Activity C3 — Print a Message for Each Item

```
foods = ["pizza", "tacos", "salad"]

for each food in foods
    print "I like " + food
end loop
```

Write the three sentences the computer will print.

____

____

____

Helper Box		
Hint	*Common Mistake*	*What to Look For*
The loop adds each food to the message.	Kids forget to include the food name.	Three full sentences.

Activity C4 — Loop Through a List With Four Items

```
materials = ["plastic", "paper", "glass", "metal"]

for each material in materials
    print material
end loop
```

Write all four materials in the order they will print.

Helper Box		
Hint	*Common Mistake*	*What to Look For*
The loop prints one item per line.	Kids reorder the list.	plastic → paper → glass → metal.

Activity C5 — Loop Through a List After a Change

```
animals = ["dog", "cat", "turtle"]
animals[1] = "parrot"

for each animal in animals
    print animal
end loop
```

Write the three items the loop will print.

____

____

____

Helper Box		
Hint	*Common Mistake*	*What to Look For*
The list is changed *before* the loop runs.	Kids use the old list.	dog → parrot → turtle.

Activity C6 — Loop Through a List and Count Items

```
games = ["Minecraft", "Roblox", "Zelda"]

count = 0

for each game in games
    count = count + 1
end loop

print count
```

What number will the computer print at the end?

Helper Box		
Hint	*Common Mistake*	*What to Look For*
The loop runs once for each item.	Kids think count starts at 1.	The total number of items.

Activity C7 — Loop Through a List and Print Only the First Letter

```
sensors = ["Air", "Light", "Noise"]

for each sensor in sensors
    print the first letter of sensor
end loop
```

Write the three letters the loop will print.

Helper Box		
Hint	*Common Mistake*	*What to Look For*
Look at the first letter of each sensor name.	Kids print the whole name.	A, L, N.

Activity C8 — Create Your Own List and Loop

Create a list of three favorite places, then write a loop that prints each one.

```
places = 

for each place in places



end loop
```

Helper Box		
Hint	*Common Mistake*	*What to Look For*
The loop should print each place.	Kids forget to use the loop variable.	A valid list and a loop that prints each item.

Activity Set D: Debugging Lists

Activity D1 — Wrong Index Number

```
animals = ["dog", "cat", "turtle"]

print animals[3]
```

Why is this code incorrect? Rewrite the code to print “cat.”

Helper Box		
Hint	*Common Mistake*	*What to Look For*
The highest index is 2.	Kids think the list has an index for “the next item.”	`print animals[1]`

Activity D2 — Missing Square Brackets

```
colors = "red", "blue", "green"
```

What is wrong with how this list is written? Rewrite the list correctly.

Helper Box		
Hint	*Common Mistake*	*What to Look For*
Lists must use square brackets.	Kids write items with commas but no brackets.	The list should be inside `[ ]`.

Activity D3 — Changing the Wrong Item

```
foods = ["pizza", "tacos", "salad"]

foods[0] = "soup"
```

If the goal was to change "salad," what is the mistake? Rewrite the code so it works correctly.

Helper Box		
Hint	*Common Mistake*	*What to Look For*
Check the index of "salad."	Kids change index 0 instead of index 2.	salad is at index 2.

Activity D4 — Missing a Comma Between Items

```
pets = ["hamster", "cat", "dog", "lizard" "bird", "fish"]
```

What is wrong with this list? Rewrite the list assignment correctly.

Helper Box		
Hint	*Common Mistake*	*What to Look For*
Check between every pair of items.	Overlooking the missing comma between `"lizard"` and `"bird"`	Add the missing comma between `"lizard"` and `"bird"`

Activity D5 — Printing the List Name Instead of an Item

```
games = ["Minecraft", "Roblox"]

print games
```

If the goal was to print only "Roblox," what is the mistake? Rewrite the code to print "Roblox."

Helper Box		
Hint	*Common Mistake*	*What to Look For*
Printing the list prints *everything.*	Kids forget to use an index.	Should be print games`[1]`.

Activity D6 — Changing an Item That Doesn't Exist

```
books = ["Dog Man", "Harry Potter"]

books[2] = "Wings of Fire"
```

Why does this code cause an error? Rewrite the code so it replaces "Harry Potter" with "Wings of Fire."

Helper Box		
Hint	*Common Mistake*	*What to Look For*
The list has only two items.	Kids try to change an index that isn't there.	Index 2 doesn't exist yet.

Activity D7 — Looping Through the Wrong List

```
fruits = ["apple", "banana", "grape"]
veggies = ["carrot", "pepper"]

for each food in veggies
    print food
end loop
```

If the goal was to print the fruits, what is the mistake? Rewrite the loop so it works correctly.

Helper Box		
Hint	*Common Mistake*	*What to Look For*
Look at which list the loop uses.	Kids mix up variable names.	The loop should use `fruits`.

Activity D8 — Adding to the Wrong List

```
names = ["Casey", "Alex"]
games = ["chess"]
games.append("Morgan")
```

The goal is to add "Morgan" to the list of names. What is the mistake? Rewrite the incorrect line.

Helper Box		
Hint	*Common Mistake*	*What to Look For*
Check which list is supposed to grow.	Changing an existing name instead of adding the new one.	`names.append("Morgan")`

Activity Set E: List Trials

Activity E1 — Two Changes Before the Loop

```
pets = ["dog", "cat", "hamster"]

pets[1] = "parrot"
pets.append("turtle")

for each pet in pets
    print pet
end loop
```

Write all four items the loop will print, in order.

Helper Box		
Hint	*Common Mistake*	*What to Look For*
The list is changed *before* the loop runs.	Kids forget the new item added at the end.	dog → parrot → hamster → turtle.

Activity E2 — Predict the Output With Numbers

```
scores = [4, 6, 8]

for each score in scores
    print score + 1
end loop
```

Write the three numbers the computer will print.

___

___

Helper Box		
Hint	*Common Mistake*	*What to Look For*
The loop prints each score *plus one.*	Kids print the original numbers.	5, 7, 9.

Activity E3 — Fix the Wrong Index

```
colors = ["red", "blue", "green"]

print colors[3]
```

Rewrite the print line so it prints "green" correctly.

Helper Box		
Hint	*Common Mistake*	*What to Look For*
The last item is at index 2.	Kids think the last item is index 3.	print `colors[2]`.

Activity E4 — Loop Through a List and Count Only Some Items

```
foods = ["pizza", "salad", "tacos", "soup"]

count = 0

for each food in foods
    if food is "soup"
        count = count + 1
    end if
end loop

print count
```

What number will the computer print?

Helper Box		
Hint	*Common Mistake*	*What to Look For*
The count increases only when the item is "soup."	Kids count all items.	Only one item matches.

Activity E5 — Change an Item, Then Print the First Letter

```
names = ["Avery", "Parker", "Robin"]

names[0] = "Skyler"

for each name in names
    print first letter of name
end loop
```

Write the three letters the loop will print.

Helper Box		
Hint	*Common Mistake*	*What to Look For*
Use the updated list.	Kids use “A” instead of “S” for the first name.	S, P, R..

Activity E6 — Add Two Items, Then Read One

```
games = ["Minecraft"]

games.append("Roblox")
games.append("Zelda")

print games[2]
```

What will the computer print?

Helper Box		
Hint	*Common Mistake*	*What to Look For*
Append adds items to the end.	Kids think the new items go at the front.	The item at index 2 is "Zelda."

Activity E7 — Debug the List and Loop

```
planets = ["Mercury", "Venus", "Earth"]

for each planet in planets
    print planets
end loop
```

This code doesn't run. What is wrong with the list, and how should it be fixed? Rewrite the incorrect line so it is correct.

Helper Box		
Hint	*Common Mistake*	*What to Look For*
Compare the loop variable with the print line.	Changing the list instead of changing what the loop prints.	Replace `print planets` with `print planet`.

Activity E8 — Create Your Own List Challenge

Create a list of four favorite songs, then write a loop that prints each one with the message “Now playing: ” before it.

Write your list:

```
songs =
```

Write your loop:

```
for each song in songs

end loop
```

Helper Box		
Hint	*Common Mistake*	*What to Look For*
Remember the square brackets, commas, and the print statement.	Kids forget the message or forget to use the loop variable.	A valid list and a loop that prints each item with the message.

CERTIFICATE OF ACHIEVEMENT

List Wizard

This certificate is awarded to ______________________________ for completing all activities in **Chapter 8: Lists** of *Think Like a Coder*.

___________________________ ___________________________

Helper Signature Date

LOGIC LAB BOOKS

Chapter 9: Boolean Logic

Combine Ideas to Decide

Introduction: Boolean Logic

You have learned a lot of clever tricks so far. Your code can repeat steps, make choices, remember things, and even work with whole lists at once. And sometimes your code needs to think a little deeper.

What if it should only move forward when **two** things are true? What if it should sound an alarm when **either** of two shapes appear? What if it should keep going if something is **not** happening?

Humans do this kind of thinking all the time:

- "I'll go outside if it's warm **and** sunny."
- "We can have pancakes if we have eggs **or** waffles if we don't."
- "I'm **not** tired... okay, maybe I am."

Computers can think this way too — they just use special logic words to do it.

In this chapter, you'll learn how to combine ideas, flip ideas, and make decisions that feel a whole lot smarter.

Mini-Lesson: AND, OR, NOT

Boolean logic gives you three new ways to think about conditions. These small words can make a *big* difference.

`and` — Both Things Must Be True

and is like a double-lock. Your code only runs when **everything** matches.

- "If the robot sees a circle **and** the path is clear, move forward."
- "I'll play outside if it's warm **and** not raining."

If even one part is false, and says "nope."

`or` — At Least One Thing is True

`or` is more relaxed. Your code runs if **any** of the conditions are true.

- "If the robot sees a triangle **`or`** a square, turn on the light."
- "We can have dessert if we have ice cream **`or`** cookies."

`or` is happy with whichever one works.

`not` — Flip the Idea Around

`not` is a switch. It takes a condition and turns it upside-down.

- "If the robot does **`not`** see a key, keep searching."
- "I'm **`not`** hungry... oh wait, maybe I am."

`not` changes true to false and false to true.

Putting Them Together

Once you know and, `or`, and `not`, you can mix them to build smarter rules and code:

- "If the list is **`not`** empty **and** the first item is red..."
- "If the robot hears a beep **`or`** the timer is done..."

This is where your code starts to feel like it's really thinking.

Activity Set A: Boolean Logic (and, or, not)

Activity A1 — Two Things at Once (and)

Read the rule and underline any situation where both parts are true.

Rule: You can go outside if it's warm **and** sunny.

Situations:

- Warm and cloudy
- Cold and sunny
- Warm and sunny
- Cold and cloudy

Helper Box		
Hint	*Common Mistake*	*What to Look For*
AND means *both* must match.	Choosing warm **or** sunny instead of warm **and** sunny.	warm and sunny

Activity A2 — Pick One (`or`)

Read the rule and underline any option that makes it true.

Rule: You can have a snack if we have apples **or** crackers.

- apples
- crackers
- bananas
- grapes

Helper Box		
Hint	*Common Mistake*	*What to Look For*
OR means *at least one* works.	Thinking both are required.	Apples, crackers.

Activity A3 — Flip It (`not`)

Read the rule and cross out the option that means you should stop reading.

Rule: Keep reading if you are **not** tired.

Options:

- tired
- not tired

Helper Box		
Hint	*Common Mistake*	*What to Look For*
NOT flips the idea.	Kids underline "tired" instead of crossing it out.	~~tired~~

Activity A4 — and **in Real Life**

Read the rule and underline the choices that follow it.

Rule: You can play the game if your room is clean **and** your homework is done.

- Room clean, homework done
- Room clean, homework not done
- Room messy, homework done

Helper Box		
Hint	*Common Mistake*	*What to Look For*
Both parts must be true.	Underlining the "clean room" option even if homework isn't done.	Only the first choice.

Activity A5 — or **in Real Life**

Read the rule and underline anything that makes it true.

Rule: You can choose a movie if it's funny **or** animated.

- Funny
- Animated
- Serious
- Boring

Helper Box		
Hint	*Common Mistake*	*What to Look For*
or means one is enough.	Underlining only one option instead of both.	funny and animated.

Activity A6 — not in Real Life

Read the rule and underline the choices that follow the rule.

Rule: You may enter if you are `not` wearing muddy shoes.

- Muddy shoes
- Clean shoes
- No shoes

Helper Box		
Hint	*Common Mistake*	*What to Look For*
not means the opposite.	Underlining "muddy shoes."	Clean shoes and No shoes.

Activity A7 — Mixed Logic

Read each rule and decide if it uses and, `or`, or `not`. Underline the correct logic word.

- You can have dessert if you finish dinner and clear your plate.
- You may pick a book if it's short or has pictures.
- You can't start the game if the controller is not connected.

Helper Box		
Hint	*Common Mistake*	*What to Look For*
Listen for the logic word.	Kids circle the wrong word because they focus on the story.	and, `or`, `not` in the correct order.

Activity A8 — Which Rule Fits?

Match each rule to the correct logic word.

You can go outside if it's not raining.	and
You can play if your chores are done and your backpack is packed.	or
You may choose a treat if we have cookies or fruit.	not

Helper Box		
Hint	*Common Mistake*	*What to Look For*
Look for the clue word in each sentence.	Mixing up and and or.	Each sentence contains the word matched to it.

Activity Set B: Fix the Rule

Activity B1 — Too Many Words

This rule works, even though it's written in a confusing way.

Rule: You may start the puzzle if you are finished with lunch and you are also done eating.

Rewrite the rule in a clearer way.

Helper Box		
Hint	*Common Mistake*	*What to Look For*
Look for repeated ideas.	Kids copy the whole sentence instead of simplifying.	Something like "Start the puzzle if you're done with lunch."

Activity B2 — Redundant `or`

This rule uses `or` in a way that makes part of it unnecessary.

Rule: You may choose a book if it has pictures or it has pictures and is short.

Rewrite the rule so it says the same thing with fewer words.

Helper Box		
Hint	*Common Mistake*	*What to Look For*
If one part already covers the other, remove the extra.	Keeping both parts.	Something like "Choose a book if it has pictures."

Activity B3 — `not` in the Wrong Place

The rule uses `not`, and it makes the sentence confusing.

Rule: You may join the game if you are not not ready.

Rewrite the rule so the meaning is clear.

____

____

Helper Box		
Hint	*Common Mistake*	*What to Look For*
Double `not` flips back to the original meaning.	Kids keep both `not`s.	Something like "Join the game if you are ready."

Activity B4 — and **Doesn't Matter**

This rule uses and. One part doesn't actually change anything.

Rule: You may have a turn if the timer beeps and the timer makes a beeping sound.

Rewrite the rule without the unnecessary part.

Helper Box		
Hint	*Common Mistake*	*What to Look For*
The second condition restates the first condition.	Kids keep both descriptions of the same event.	Something like "Have a turn if the timer beeps."

Activity B5 — `or` Makes Everything True

This rule uses `or` in a way that makes the rule always true.

Rule: You may recycle an item if it is plastic or not plastic.

Rewrite the rule so it makes sense — pick a different material.

Helper Box		
Hint	*Common Mistake*	*What to Look For*
`or` shouldn't include opposites.	Kids rewrite the same rule.	A new rule using `or` that makes sense (e.g., "if it's metal or glass").

Activity B6 — Too Many `nots`

This rule uses `not` twice, which makes it confusing.

Rule: You may not line up if you are not finished cleaning.

Rewrite the rule so it says the same thing in a **clearer way**.

Helper Box		
Hint	*Common Mistake*	*What to Look For*
Two `nots` can flip the meaning around.	Kids keep both `nots` in the new rule.	A simple rule with no `nots`, like "You may line up when you are finished cleaning."

Activity B7 — Mixed-Up Logic

Each rule uses the wrong logic word.

1. You may start the craft if the glue is open or the scissors are ready. **(Should require both.)**
2. You may choose a game if it's short and fun. **(Should allow either.)**
3. You may begin if you are not not paying attention. **(Should be the simple version.)**

Rewrite each rule using the correct logic word.

1.

2.

3.

Helper Box		
Hint	*Common Mistake*	*What to Look For*
Think about what the rule *should* mean.	Kids only change one word.	1 – and, 2 – or , 3 – remove double not

Activity B8 — Fix the Whole Thing

This rule is confusing, repetitive, and uses the wrong logic word.

Rule: You may choose a snack if you finish your chores or you finish your chores and your room is clean.

Rewrite the rule so it is clear and correct.

____

____

Helper Box		
Hint	*Common Mistake*	*What to Look For*
Look for the real requirement.	Keeping OR when AND is needed.	"Choose a snack if you finish your chores and your room is clean."

Activity Set C: Break Down the Logic

Activity C1 — Find the Conditions

Read the rule. Write the **first condition** on one line and the **second condition** on the next line.

Rule: You may play outside if it is warm and the grass is dry.

Condition 1:

Condition 2:

Helper Box		
Hint	*Common Mistake*	*What to Look For*
Look for the two ideas connected by and.	Kids rewrite the whole sentence instead of splitting it.	"It is warm" and "the grass is dry."

Activity C2 — Find the Conditions

Read the rule. Write the **two conditions** that make the rule true.

Rule: "You may choose a craft if it uses paint or it uses stickers."

Condition 1:

Condition 2:

Helper Box		
Hint	*Common Mistake*	*What to Look For*
or connects two options.	Kids write only one choice.	"Uses paint" and "uses stickers."

Activity C3 — What Is not Allowed?

Read the rule. Write the thing that is **not** allowed.

Rule: You may water the garden if you are not wasting water.

Not allowed:

Helper Box		
Hint	*Common Mistake*	*What to Look For*
not flips the idea.	Kids write "wasting water" without understanding why.	"wasting water. "

Activity C4 — Break the Rule Into Three Parts

This rule has **three** conditions. Write each condition on its own line.

Rule: You may start the game if you are ready and the controller is connected and the timer has started.

Condition 1:

____

____

Condition 2:

____

____

Condition 3:

____

____

Helper Box		
Hint	*Common Mistake*	*What to Look For*
Look for each and.	Combining two parts into one.	Ready / controller connected / timer started.

Activity C5 — Which Part Makes It True?

Read the rule and the situation. Circle the condition that makes the situation true.

Rule: You may choose a snack if it is crunchy or sweet.

Situation: The snack is sweet.

crunchy sweet

Helper Box		
Hint	*Common Mistake*	*What to Look For*
`or means at least one part needs to match.`	Kids circle both.	"sweet"

Activity C6 — Which Part Makes It False?

Read the rule and the situation. Circle the condition that **does not** match the rule.

Rule: You may join the activity if you are calm and listening.

Situation: You are calm. You are not listening.

calm listening

Helper Box		
Hint	*Common Mistake*	*What to Look For*
and needs both parts to be true.	Kids circle "calm" because they see it first.	"listening."

Activity C7 — Break Down a Mixed Rule

This rule uses and and `not`. Write each condition on its own line.

Rule: You may choose a book if it is short and not scary.

Condition 1:

Condition 2:

Helper Box		
Hint	*Common Mistake*	*What to Look For*
`not` belongs to the second condition.	Kids split "not scary" into two parts.	"Short" and "not scary. "

Activity C8 — Explain the Rule

Read the rule and the situation. Decide whether the situation makes the rule **true** or **false** and write a short explanation of **why**.

Rule: You may pick a game if it is new or you haven't played it before.

Situation: You haven't played the game before.

Helper Box		
Hint	*Common Mistake*	*What to Look For*
`or` means at least one true part is enough.	Kids restate the rule instead of explaining.	Something like "I haven't played it before, so the rule is true. "

Activity Set D: Debug the Conditions

Activity D1 — Which Situation Follows the Rule?

Rule: You may choose a craft if it uses paint and has enough space.

Circle the situation that follows the rule, then explain why the other situation does not follow the rule.

Situation A	Situation B
The craft uses paint and has enough space.	The craft uses paint and does not have enough space.

Helper Box		
Hint	*Common Mistake*	*What to Look For*
and means both conditions need to be true.	Kids choose A without explaining why B fails.	Both conditions are true

Activity D2 — Which Situation Follows the Rule?

Rule: You may choose a snack if it is salty or crunchy.

Circle the situation that follows the rule, then explain why the other situation does not follow the rule.

Situation A	Situation B
The snack is not salty and not crunchy.	The snack is salty and not crunchy.

Helper Box		
Hint	*Common Mistake*	*What to Look For*
or means at least one condition needs to be true.	Kids think both conditions must be true.	Salty and not crunchy.

Activity D3 — Which Situation Follows the Rule?

Rule: You may join the group if you are not running.

Circle the situation that follows the rule, then explain why the other situation does not follow the rule.

Situation A	Situation B
You are walking.	You are running.

Helper Box		
Hint	*Common Mistake*	*What to Look For*
`not` running is the condition.	Kids confuse walking with running.	Walking meets the condition and follows the rule.

Activity D4 — Which Condition was Checked Incorrectly?

Rule: The robot may start moving if its battery is charged and its sensors are working.

Situation: The battery is charged. The sensors are not working.

Outcome: The robot started moving.

Write the condition that was checked incorrectly.

____

____

Helper Box		
Hint	*Common Mistake*	*What to Look For*
and requires both conditions to be true.	Kids write "not working" instead of naming the condition.	"sensors working."

Activity D5 — Which Condition was Checked Incorrectly?

Rule: You may choose a book if it is short or has pictures.

Situation: The book is long and it has pictures.

Outcome: You were not allowed to choose it.

Write the condition that was checked incorrectly.

____

____

Helper Box		
Hint	*Common Mistake*	*What to Look For*
or means one true condition is enough.	Kids say "pictures" without explaining the logic.	The "has pictures" condition.

Activity D6 — Which Condition was Checked Incorrectly?

Rule: You may line up if you are calm and not talking.

Situation: You are calm and quiet.

Outcome: You were told you may not line up.

Write the condition that was checked incorrectly.

____

____

Helper Box		
Hint	*Common Mistake*	*What to Look For*
not talking means quiet is okay.	Kids say "calm" because it appears first.	The "not talking" condition.

Activity D7 — List the Ignored Conditions

Rule: You may choose a game if it is new and clean and not missing pieces.

Situation: The game is old, dirty, and not missing pieces.

Outcome: You were allowed to choose it.

List all the conditions that were ignored or misunderstood:

____

____

____

Helper Box		
Hint	*Common Mistake*	*What to Look For*
Check each condition separately.	Kids only list one condition.	"new" and "clean"

Activity D8 — List the Ignored Conditions

Rule: You may start the project if your supplies are ready and your partner is ready and the instructions are open.

Situation: Supplies ready, partner not ready, instructions closed.

Outcome: You were allowed to start.

List all the conditions that were ignored or misunderstood:

Helper Box		
Hint	*Common Mistake*	*What to Look For*
Every condition must be true for and.	Kids only list the first incorrect condition.	"partner is ready" and "instructions are open"

Activity Set E: Create Your Own Logic

Activity E1 — Write Your Own AND Rule

Write a rule that uses and to connect two conditions.

You may ______________________________

if ______________________________

and ______________________________

Helper Box		
Hint	*Common Mistake*	*What to Look For*
Both conditions must be true.	Kids forget to include two conditions.	A rule with two clear conditions joined by and.

Activity E2 — Write Your Own or Rule

Write a rule that uses `or` to connect two conditions.

Helper Box		
Hint	*Common Mistake*	*What to Look For*
At least one condition needs to be true.	Kids may write a rule using and.	Two conditions joined by `or`.

Activity E3 — Write a Rule Using `not`

Write a rule that uses `not` to flip a condition.

Helper Box		
Hint	*Common Mistake*	*What to Look For*
`not` applies to one condition.	Kids put `not` in front of the whole sentence.	A single condition flipped with `not`.

Activity E4 — Create Two Situations

Rule: You may choose a game if it is short or new.

Write two situations for the rule: one that follows the rule and one that does not follow it.

Situation that follows the rule:

Situation that does not follow the rule:

Helper Box		
Hint	*Common Mistake*	*What to Look For*
One situation should match at least one condition.	Kids write two matching situations.	One true situation, one false situation.

Activity E5 — Make a Mixed Rule

Write a rule that uses `not` with and.

Helper Box		
Hint	*Common Mistake*	*What to Look For*
`not` belongs to one condition.	Kids put `not` in the wrong place.	A rule like "if it is ________ `and not` ________."

Activity E6 — Explain Your Rule

Choose one of your rules from Activities E1–E5. Explain why the rule works the way it does.

Helper Box		
Hint	*Common Mistake*	*What to Look For*
Explain how the conditions affect the rule.	Kids restate the rule instead of explaining it.	A sentence starting with something like "It works because..."

Activity E7 — Create a Multi-Condition Rule

Write a rule with three conditions using any combination of and, or, and not.

Helper Box		
Hint	*Common Mistake*	*What to Look For*
Think about how the conditions work together.	Kids may write fewer than three conditions or leave them disconnected.	A rule with three conditions and at least two appropriately used logic operators.

Activity E8 — Write a Rule and a Matching Situation

Write your own rule using any Boolean words. Then write a situation that follows your rule.

Your rule:

____

____

____

Situation that follows the rule:

____

____

____

Helper Box		
Hint	*Common Mistake*	*What to Look For*
The situation must match the conditions exactly.	Kids write a situation that only matches part of the rule.	A situation that satisfies all required conditions.

CERTIFICATE OF ACHIEVEMENT

True/False Titan

This certificate is awarded to ____________________________ for completing all activities in **Chapter 9: Boolean Logic** of *Think Like a Coder*.

________________________ Helper Signature

________________________ Date

LOGIC LAB BOOKS

Chapter 10: Power-Ups

Coding Super Moves

Introduction: Power-Ups

You've learned all the big building blocks of coding — loops, lists, conditions, and functions. Now it's time to combine them.

This chapter is your **Power-Up Zone**. You'll learn how coding ideas become *way* more powerful when you use them together:

- loops inside loops
- loops that search
- loops that count
- loops that build new lists
- loops that call functions
- loops that use multi-condition logic

These are the same patterns real programmers use in games, apps, robots, and everything else.

You're not learning new tools — you're learning how to **combine** the tools you already have.

Get ready to level up your coding superpowers.

Mini-Lesson: Combine Concepts to Level Up

Nested Loops

A nested loop is a loop *inside* another loop. It's how you make rows, columns, grids, and repeated patterns.

Example:

```
rows = 2
columns = 3
for row from 1 to rows
     for col from 1 to columns
          print "*"
     end loop
end loop
```

Result:

```
***
***
```

The outer loop repeats the inner loop.

Searching a List With a Loop

A loop can check each item in a list until it finds what you're looking for.

```
pets = ["dog", "cat", "hamster"]
for each pet in pets
     if pet is "hamster"
          print "Found it!"
     end if
end loop
```

Result: Found it!

The loop checks each item and prints the message whenever the condition is true.

Counting With a Loop

Loops can update a variable each time they run — like counting or adding.

Example:

```
colors = ["red", "blue", "red"]
count = 0
for each color in colors
     if color is "red"
          count = count + 1
     end if
end loop
print count
```

Result: 2

The loop keeps track of how many items match the condition.

Filtering a List

A loop can build a new list by keeping only the items that match a condition.

Example:

```
foods = ["pizza", "salad", "tacos", "salad"]
healthy = []
for each food in foods
     if food is "salad"
          add food to healthy
     end if
end loop
print healthy
```

Result: `["salad", "salad"]`

This is how computers sort and organize information.

Loop + Function Power-Up

A loop can call a function again and again — one time for each item, and it can *pass* the item to the function.

Functions Can Receive Information

A function can receive information to use while it runs. The name inside the parentheses is a temporary variable called a parameter.

```
function greet(animal)
    print "Hello, " + animal
end function
```

When you write `greet("cat")`, the value `"cat"` is stored in the variable named `animal` while the function runs. This lets the same function work with many different items.

Example:

```
animals = ["cat", "dog", "bird"]
function greet(animal)
     print "Hello, " + animal
end function

for each ani in animals
     greet(ani)
end loop
```

Result:

```
Hello, cat
Hello, dog
Hello, bird
```

This is how programs reuse code to do big jobs.

Multi-Condition Logic in Loops

Loops can use more than one condition at a time.

Example:

```
books = ["short funny", "long serious", "short silly", "short
serious"]
for each book in books
     if book has "short" and book is not "serious"
          print book
     end if
end loop
```

Result:

```
short funny
short silly
```

This is where loops and Boolean logic team up.

Activity Set A: Loop Power-Ups

Activity A1 — Loop With a Condition

```
foods = ["pizza", "salad", "tacos"]

for each food in foods
    if food is "salad"
        print "Healthy choice!"
    end if
end loop
```

What will the computer print?

Helper Box		
Hint	*Common Mistake*	*What to Look For*
The message prints only when the item is “salad.”	Kids print the message for every item.	Only one line of output.

Activity A2 — Loop That Searches for Something

```
pets = ["dog", "cat", "hamster"]

for each pet in pets
    if pet is "hamster"
        print "Found it!"
    end if
end loop
```

What will the computer print?

Helper Box		
Hint	*Common Mistake*	*What to Look For*
The message prints only when the item matches.	Kids print all three items.	One "Found it!" message.

Activity A3 — Loop That Stops on a Match

```
names = ["Jamie", "Rowan", "Cameron"]

for each name in names
    if name is "Rowan"
        print "Here!"
        stop loop
    end if
end loop
```

Which item in the list makes the loop print "Here!"?

Helper Box		
Hint	*Common Mistake*	*What to Look For*
The loop stops immediately after the matching name.	Kids continue checking names after the match.	The second item.

Activity A4 — Loop That Adds Numbers

```
numbers = [2, 3, 5]

total = 0

for each num in numbers
    total = total + num
end loop

print total
```

What number will the computer print?

____

____

____

Helper Box		
Hint	*Common Mistake*	*What to Look For*
Add each number to total.	Kids forget that total starts at 0.	The sum of all three numbers.

Activity A5 — Loop That Counts Matches

```
materials = ["plastic", "paper", "plastic", "metal"]

count = 0

for each material in materials
    if material is "plastic"
        count = count + 1
    end if
end loop

print count
```

What number will the computer print?

Helper Box		
Hint	*Common Mistake*	*What to Look For*
Count how many times "plastic" appears.	Kids count all items.	Two matches.

Activity A6 — Loop That Builds a Pattern

```
repeat 3 times
    print "* * *"
end loop
```

Write exactly what the computer will print.

Helper Box		
Hint	*Common Mistake*	*What to Look For*
The same pattern prints three times.	Kids change the pattern or print it fewer times.	Three identical lines.

Activity A7 — Loop With a Changing Variable

```
num = 1

for step from 1 to 3
    num = num + 2
end loop

print num
```

What number will the computer print?

____

____

Helper Box		
Hint	*Common Mistake*	*What to Look For*
num increases by 2 each time the loop runs.	Kids think num resets each time.	num starts at 1 and increases by 2 three times, ending with a value of 7.

Activity A8 — Create Your Own Loop With a Condition

Create a list of three favorite animals, then write a loop that prints "I love " + the animal only when the animal is your favorite of the three.

Write your list:

```
animals =
```

Write your loop:

```
for each animal in animals



end loop
```

Helper Box		
Hint	*Common Mistake*	*What to Look For*
The loop should print the message only for one chosen animal.	Kids print the message for every item.	A correct list and a condition inside the loop.

Activity Set B: List Power-Ups

Activity B1 — Pick the Items That Match

Here is a list. Underline the items in the list that match the condition.

```
colors = ["red", "blue", "green", "red"]
```

Condition: The color is "red".

Helper Box		
Hint	*Common Mistake*	*What to Look For*
Check each item one at a time.	Kids stop after finding the first match.	Both red items.

Activity B2 — Pick the Items That Do NOT Match

Here is a list. Underline all the items that match the condition.

```
pets = ["dog", "cat", "fish", "cat"]
```

Condition: The pet is not "cat".

Helper Box		
Hint	*Common Mistake*	*What to Look For*
`not` flips the condition.	Kids include “cat” by accident.	dog and fish

Activity B3 — Pick Items That Match One of Two Conditions

Here is a list. Underline all the items that match the condition.

```
snacks = ["chips", "apple", "cookie", "carrot"]
```

Condition: The snack is "chips" or "cookie".

Helper Box		
Hint	*Common Mistake*	*What to Look For*
or means either condition is okay.	Kids include only one of the two.	chips and cookie

Activity B4 — Pick Items That Match BOTH Conditions

Here is a list of toys. Underline the items that match the condition.

```
toys = ["small car", "big car", "small doll", "big puzzle"]
```

Condition: The toy is "small" and "car".

Helper Box		
Hint	*Common Mistake*	*What to Look For*
Both conditions must be true.	Kids pick “small doll” because it’s small.	small car

Activity B5 — Pick Items That Match a Mixed Condition

Here is a list. Write the items that match the condition.

```
books = ["short funny", "long funny", "short serious", "long serious", "short nonfiction"]
```

Condition: The book is "short" and not "serious".

Helper Box		
Hint	*Common Mistake*	*What to Look For*
Break the condition into two parts.	Kids forget the `not`.	short funny short nonfiction

Activity B6 — Count the Items That Match

Here is a list.

```
numbers = [3, 7, 3, 3, 2, 3]
```

Condition: The number is 3.

How many items match the condition?

Helper Box		
Hint	*Common Mistake*	*What to Look For*
Count each match.	Kids count only the first or last one.	4 matches.

Activity B7 — Count Items That Match One of Two Conditions

Here is a list.

```
animals = ["cat", "dog", "bird", "cat", "fish"]
```

Condition: The animal is "cat" or "fish".

How many items match the condition?

Helper Box		
Hint	*Common Mistake*	*What to Look For*
`or` means either condition is okay.	Kids count only the cats.	3 matches.

Activity B8 — Create Your Own List Filter

Create a list of five things. Then write a condition that picks only some of them. Finally, write the items that match your condition.

Your list:

Your condition:

Items that match:

Helper Box		
Hint	*Common Mistake*	*What to Look For*
Make your condition clear and specific.	Kids write a condition that matches everything.	A condition that selects only some items.

Activity Set C: Loops + Lists Power-Ups

Example

Here is a list with a loop.

```
numbers = [12, 5, 2]
count = 20
for each num in numbers
    count = count - num
end loop
```

Write the value of `count` **after each step:**

After `num = 12`: count = 8

After `num = 5`: count = 3

After `num = 2`: count = 1

Activity C1 — Track the Variable After Each Step

```
numbers = [2, 4, 6]
sum = 0
for each num in numbers
     sum = sum + num
end loop
```

Write the value of `sum` after each step:

After `num = 2`:

After `num = 4`:

After `num = 6`:

Helper Box		
Hint	*Common Mistake*	*What to Look For*
Update `sum` one step at a time.	Kids may jump straight to the final answer.	2 → 6 → 12.

Activity C2 — Track a List as It Grows

```
pets = ["dog", "cat", "fish"]
favorites = []
for each pet in pets
     add pet to favorites
end loop
```

Write what `favorites` looks like after each step:

After `pet = "dog"`:

____

____

After `pet = "cat"`:

____

____

After `pet = "fish"`:

____

____

Helper Box		
Hint	*Common Mistake*	*What to Look For*
Add one item at a time.	Kids write only the final list.	`["dog"]` → `["dog","cat"]` → `["dog","cat","fish"]`.

Activity C3 — When Does the Loop Stop?

```
names = ["Jamie", "Rowan", "Cameron"]
for each name in names
     if name is "Rowan"
          stop loop
     end if
end loop
```

Write the items the loop checks before stopping:

____

____

Helper Box		
Hint	*Common Mistake*	*What to Look For*
The loop stops as soon as the condition is true.	Kids include "Cameron."	Jamie, Rowan.

Activity C4 — Track Two Variables at Once

```
numbers = [1, 3, 5]
sum = 0, count = 0
for each num in numbers
     sum = sum + num
     count = count + 1
end loop
```

Write the value of `sum` and `count` after each step:

After `num = 1`:	`sum =`	`count =`
After `num = 3`:	`sum =`	`count =`
After `num = 5`:	`sum =`	`count =`

Helper Box		
Hint	*Common Mistake*	*What to Look For*
Update both variables each time.	Kids update only one variable.	sum = 1, 4, 9. count = 1, 2, 3.

Activity C5 — Transform Each Item in a List

```
words = ["hi", "bye", "yo"]
new_words = []
for each word in words
     add word + "!" to new_words
end loop
```

Write what `new_words` looks like after each step:

After `word = "hi"`:

After `word = "bye"`:

After `word = "yo"`:

Helper Box		
Hint	*Common Mistake*	*What to Look For*
Add the transformed item, not the original.	Kids forget the "!".	`["hi!"]` → `["hi!", "bye!"]` → `["hi!", "bye!", "yo!"]`.

Activity C6 — Count AND Build a List

```
colors = ["red", "blue", "red", "green"]
count = 0
reds = []
for each color in colors
     if color is "red"
          count = count + 1
          add color to reds
     end if
end loop
```

Write the value of count and the list reds after each loop iteration:

After first "red":	count =	reds =
After "blue":	count =	reds =
After second "red":	count =	reds =
After "green":	count =	reds =

Helper Box		
Hint	*Common Mistake*	*What to Look For*
Update both the variable and the list.	Kids update only the variable or the list.	count: 1,1,2,2. reds: ["red"],["red"],["red"," red"],["red","red"]

Activity C7 — Use a "Found" Flag

```
pets = ["dog", "cat", "hamster"]
found = false
for each pet in pets
     if pet is "cat"
          found = true
     end if
end loop
```

Write the value of `found` after each step:

After `"dog"`:

____

____

After `"cat"`:

____

____

After `"hamster"`:

____

____

Helper Box		
Hint	*Common Mistake*	*What to Look For*
The flag stays true once it becomes true.	Kids flip it back to false.	false → true → true.

Activity C8 — Create Your Own Step-By-Step Trace

Create a list of four items. Write a loop that changes a variable or builds a new list. Then write what happens after each step.

Your list:

Your loop:

After step 1:

After step 2:

After step 3:

After step 4:

Helper Box		
Hint	*Common Mistake*	*What to Look For*
Show the loop running one item at a time.	Kids write only the final answer.	A clear step-by-step trace.

Activity Set D: Debug Multi-Concept Rules

Activity D1 — The Loop Never Runs

This code is trying to print each of the foods in the list.

```
foods = ["pizza", "salad", "tacos"]
items = []
for each food in items
    print food
end loop
```

Outcome: Nothing prints.

What is the mistake?

____

____

____

Helper Box		
Hint	*Common Mistake*	*What to Look For*
Compare the list in the loop with the list containing the foods.	Changing the print instruction.	Change `items` to `foods`.

Activity D2 — The Function Never Gets Called

This code is trying to cheer for each number greater than 5.

```
function cheer(number)
    print "Yay for " + number
end function
numbers = [3, 5, 7]
for each num in numbers
    if num is greater than 10
        cheer(num)
    end if
end loop
```

Outcome: Nothing prints.

What is the mistake?

____

____

Helper Box		
Hint	*Common Mistake*	*What to Look For*
Compare the condition with the intended numbers.	Kids think the function is broken.	The condition should select numbers greater than 5.

Activity D3 — The Wrong Function Gets Called

This code is trying to cheer for dogs.

```
function greet(ani)
    print "Hello, " + ani
end function
function cheer(ani)
    print "Yay, " + ani
end function
animals = ["cat", "dog"]
for each animal in animals
    if animal is "dog"
        greet(animal)
    end if
end loop
```

Outcome: The loop prints "Hello, dog" instead of cheering.

What is the mistake?

___

___

Helper Box		
Hint	*Common Mistake*	*What to Look For*
Look at which function is called.	Kids think the condition is wrong.	`greet` was used instead of `cheer`.

Activity D4 — The Loop Stops Too Early

This code should print each name and then stop immediately after printing Dakota.

```
names = ["Reese", "Dakota", "Rory"]
for each name in names
     if name is "Reese"
          stop loop
     end if
     print name
end loop
```

Outcome: Reese is not printed.

What is the mistake?

____

____

____

Helper Box		
Hint	*Common Mistake*	*What to Look For*
Compare the order of print name and stop loop.	Kids think the condition is wrong.	The loop stops before Reese is printed.

Activity D5 — The List Gets the Wrong Items

This code is trying to find "red" and add it to the favorites.

```
colors = ["red", "blue", "green"]
favorites = []
for each color in colors
    if color is "red"
        add "blue" to favorites
    end if
end loop
```

Outcome: `favorites = ["blue"]`

What is the mistake?

___

___

___

Helper Box		
Hint	*Common Mistake*	*What to Look For*
Look at what gets added.	Kids think the condition is wrong.	The wrong item is added.

Activity D6 — The Condition and Update Are Wrong

The code below is trying to print the numbers in the list that are greater than 4 and then print the sum of those numbers.

```
numbers = [2, 7, 9, 4, 1, 6, 5]
total = 0
for each num in numbers
     if num is even
          print num
          total = total + 1
     end if
end loop
print total
```

Outcome: The program prints:

```
2
4
6
3
```

List the mistakes.

____

____

____

Helper Box		
Hint	*Common Mistake*	*What to Look For*
Compare both the condition and the update with the stated goal.	Fixing only one of the two errors.	even instead of `greater than 4`. `total` is increased by 1 instead of `num`.

Activity D7 — The Loop Doesn't Work at All

The code is trying to call `greet()` for every pet that is a cat or a hamster.

```
pets = ["dog" "cat" "hamster"]
function greet(pet)
     print "Hello, " + pet
end function
for each pet in pets
     if pet is "cat" and not "hamster"
          greet(cat)
     end if
end loop
```

Outcome: Nothing prints. The program throws an error about a missing comma and an undefined variable.

List all three mistakes:

1. ____________________

2. ____________________

____

3. ____________________

____

Helper Box		
Hint	*Common Mistake*	*What to Look For*
Check the list, the condition, and the function call.	Kids find only the logic error.	Missing commas, incorrect condition, wrong variable passed to `greet()` function

Activity D8 — The Loop Prints the Wrong Things

The code is trying to call `celebrate(res)` when `res` includes "winner" and does not include "injured", and call `cheer(res)` when `res` is "runner" or "helper".

```
results = ["winner", "winner injured", "runner", "helper"]
function celebrate(player)
     print "Celebrate " + player
end function
function cheer(player)
     print "Cheer for " + player
end function
for each res in results
     if res has "winner" or res is missing "injured"
          celebrate(res)
     end if
     if res is "runner" and res is "helper"
          cheer(res)
     end if
end loop
```

Outcome: Every result is celebrated. Nothing is cheered.

List both mistakes:

1. ______________________________

2. ______________________________

Helper Box		
Hint	*Common Mistake*	*What to Look For*
Check both logic words.	Fixing only one condition.	Change or to and, then change and to or.

Activity Set E: Create Your Own Power-Up

Activity E1 — Create a Loop That Searches a List

Create a list of four items. Write a loop that prints a message only when the item matches your condition.

Your list: ____________________

Your condition: ____________________

Your loop: ____________________

____

____

____

____

Helper Box		
Hint	*Common Mistake*	*What to Look For*
Use `if` inside the loop.	Kids print every item.	A loop that checks the condition before printing.

Activity E2 — Create a Loop That Counts Matches

Create a list of five items. Choose one item to count. Write a loop that counts how many times it appears.

Your list:

Item to count:

Your loop:

Final count:

Helper Box		
Hint	*Common Mistake*	*What to Look For*
Start count at 0.	Kids forget to update count.	`count = count + 1` inside the condition.

Activity E3 — Create a Loop That Builds a New List

Create a list of five items. Write a loop that adds only the items that match your condition to a new list.

Your list:

Your condition:

Your loop:

New list:

Helper Box		
Hint	*Common Mistake*	*What to Look For*
Add only matching items.	Kids add everything.	A new list with only the correct items.

Activity E4 — Create a Loop That Calls a Function

Write a function that prints a message about an item. Then create a list and write a loop that calls the function for each item.

Your function:

Your list:

Your loop:

Helper Box		
Hint	*Common Mistake*	*What to Look For*
The function runs once per item.	Kids forget to call the function.	A loop that calls the function correctly.

Activity E5 — Create a Loop With a Mixed Condition

Create a list of four items. Write a loop that prints something only when two conditions are true.

Your list:

Condition 1:

Condition 2:

Your loop:

Helper Box		
Hint	*Common Mistake*	*What to Look For*
Join the two conditions with and.	Kids check only one condition.	Both conditions inside the "if."

Activity E6 — Create a Loop That Transforms Items

Create a list of three words. Write a loop that adds a changed version of each word to a new list (add "!", add "super ", add "_done", etc.).

Your list:

Your transformation:

Your loop:

New list:

Helper Box		
Hint	*Common Mistake*	*What to Look For*
Add the transformed item, not the original.	Kids forget to apply the transformation.	A new list with changed items.

Activity E7 — Create a Loop That Stops Early

Create a list of five items. Choose one item that should make the loop stop. Write a loop that prints each item before the stop item, then stops without printing the stop item.

Your list:

Stop item:

Your loop:

Helper Box		
Hint	*Common Mistake*	*What to Look For*
Use stop loop inside the condition.	Kids print the stop item too.	Loop stops immediately when the condition is true.

Activity E8 — Create Your Own Power-Up Algorithm

Write a mini-algorithm that uses at least three of these:

- a list
- a loop
- a condition
- a function
- a new list
- a variable that changes
- a stop loop
- multi-condition logic

Your idea:

Your code:

What your code does:

Helper Box		
Hint	*Common Mistake*	*What to Look For*
Start simple, then add power ups.	Kids forget one of the required parts.	A clear, multi concept algorithm that works.

CERTIFICATE OF ACHIEVEMENT

Power-Up Prodigy

This certificate is awarded to ______________________________ for completing all activities in **Chapter 10: Power-Ups** of *Think Like a Coder*.

___________________________ Helper Signature

___________________________ Date

LOGIC LAB BOOKS

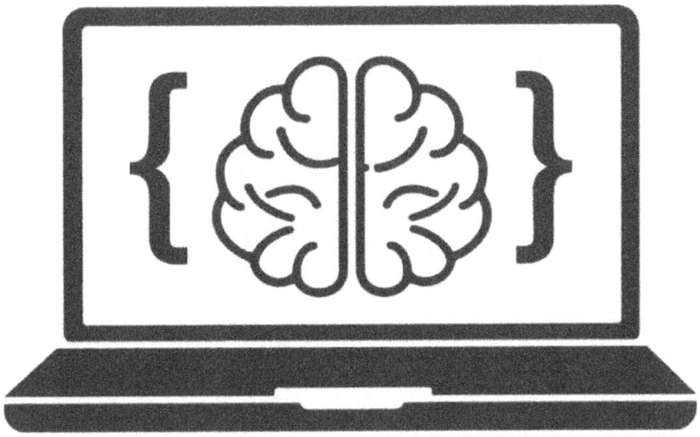

CERTIFICATE OF ACHIEVEMENT

Think Like a Coder Graduate

This certificate is awarded to ______________________________ for completing all activities in *Think Like a Coder.*

___________________________ Helper Signature

___________________________ Date

LOGIC LAB BOOKS

Lisa is a lifelong learner, maker, and STEM enthusiast who believes every child can learn to think like a programmer — long before they ever touch a computer. She creates screen-free, kid-friendly workbooks that build logic, problem-solving, and confidence through hands-on activities.

When she's not designing educational resources, Lisa spends her time caring for her alpacas, exploring creative projects, and helping kids discover the joy of learning. **Logic Lab Books** is her way of making STEM accessible, engaging, and fun for kids and families everywhere.

Logic Lab Books

Elverta, California

www.logiclabbooks.com

Inspiring young thinkers to develop a passion for STEM.

Printed in the United States of America

Fonts and Licensing

Handwriting Font — Paulo Bittencourt
Designed by Paulo Bittencourt, Licensed via 1001 Fonts for personal and commercial use
https://www.1001fonts.com/paulo-bittencourt-font.html

Monospace Font — Source Code Pro
Designed by Paul D. Hunt for Adobe, Licensed under the SIL Open Font License (OFL)
https://www.1001fonts.com/source-code-pro-font.html

Logo Font — Space Mono
Designed by Colophon Foundry, Licensed under the SIL Open Font License (OFL)
https://www.1001fonts.com/space-mono-font.html

Illustrations and Design

All icons, illustrations, and layout design created by Lisa Hargreaves. Logic Lab Books logo designed by Lisa Hargreaves.

www.ingramcontent.com/pod-product-compliance
Lightning Source LLC
LaVergne TN
LVHW070906210926
848482LV00017B/1007
* 9 7 9 8 9 9 6 7 4 7 0 0 9 *